KV-575-689

HANDBOOK OF PRECISION ENGINEERING

Edited by A. Davidson

Volume 6 Mechanical Design Applications

MACMILLAN

SBN 333 11825 1

First published in English by
THE MACMILLAN PRESS LTD
London and Basingstoke
Associated companies in New York Toronto Melbourne
Dublin Johannesburg and Madras

Trademarks of N. V. Philips' Gloeilampenfabrieken

Printed in Great Britain by The Whitefriars Press Ltd., London and Tonbridge

D
681
HAN

Foreword

During the last twenty or thirty years, precision engineering, albeit not a separate field such as, say, naval architecture or aeronautics, has nevertheless emerged as a technology in its own right, involving methods and concepts different from those of conventional engineering.

No clear definition of what is meant by precision engineering can be given. The view that this covers all those sectors of technical activity in which artificial aids are employed as extensions of the senses can no longer be upheld.

The question of what is to be regarded as a precision mechanism can best be settled by summarizing a number of devices which qualify for this description. Generally, they are small devices ranging from the unique to the mass produced.

The term precision mechanism may well be a misnomer as applied, say, to a watch, since an ordinary watch, although small, is a relatively crudely toleranced instrument, whereas the drive of an observatory telescope, a massive piece of equipment, is constructed to operate with the utmost precision. Examples of precision engineering in the sense in which the term is employed in this handbook are: telecommunications equipment; optical apparatus such as cameras, projectors, but also such instruments as microscopes, comparators, etc.; office machines such as typewriters, accounting machines, calculating machines, etc.; technical construction kits, marketed as toys, in view of the often ingenious constructions and highly sophisticated tool designs employed; electric shavers, electronic equipment for use in the home, etc. Many other examples could be added to this list.

Both in the precision engineering industry and in education, there is a need for a handbook providing ready access to all the different subjects involved in this branch of engineering. Although in most cases these subjects are not new in themselves, there is much to be gained from a review of them which places special emphasis on those aspects which have a specific bearing on precision mechanisms and which provides information of possible use to designers, manufacturers and users of products in this category.

The authors, specialists in their field, do not claim either to have covered every aspect of the subject dealt with in their chapter, or to have included all the details in the limited space available. What they have tried to do is to describe the essentials as thoroughly as possible, whilst providing an extensive list of references to the international literature as a source of detail.

The handbook has been divided into twelve parts. Part 1 deals with the general principles upon which the design of a product should be based, whilst Part 2 discusses the materials required. They contain both theory and practical information of interest to the designer and the engineer. Methods and technical processes of production are discussed in Parts 3 to 5 and 8 to 10. Here, production engineers will find full information on production methods and the machines involved.

The remaining parts (6, 7, 11 and 12) cover the construction of precision engineering products and equipment, together with components for same. Designers, manufacturers and users will all find useful data in them.

This handbook is published as a sequel to the first two parts of the *Handbook of Precision Engineering*, produced as long ago as 1957 by the Cosmos Publishing Company.

The authors take this opportunity to thank all those who have assisted in its production.

Eindhoven, 1968   A. Davidson

Contents

Chapter 3 Optical Engineering Designs 274

Introduction

A. DAVIDSON

The construction of precision engineering products differs from the conventional engineering of machines, steel structures and so on, not only because it involves a particular design philosophy, as emerges very clearly in mass production projects, but also because of the problems it presents in job production and batch production processes. Tolerances, and the proper choice of material, are of the utmost importance.

The greater part of the present volume is devoted to descriptions of structural elements. Although essential information concerning permanent joints has already been given in Volume 5, all the possible forms are described here for the sake of completeness. Many of the detachable fastenings have been standardized. Information concerning them is to be found in Part 1, as well as in the present volume. Guides, bearings, etc., although mentioned in various other parts of the Handbook, are nevertheless discussed at some length. The same applies to gear wheels and gear trains. Other mechanical components, such as locking devices, springs, couplings, friction clutches and so on, are reviewed in detail. Diagrams are employed very extensively throughout the first chapter, because a pictorial survey explains matters more clearly than any description. The captions merely clarify the details.

Chapter 2 contains examples of practical applications in precision engineering. Since it is, of course, beyond the scope of this volume to include descriptions of all possible structures, the examples given are confined to two specific branches of precision engineering, namely cinematograph equipment and sewing machines, although it is shown that similar structures are employed in other kinds of equipment. Specialized structures are used in optical apparatus, the basic elements of which are discussed in Chapter 3. For a description of the apparatus itself, see Volume 11. Because very little information on this subject is available in the literature, references to one or two handbooks of optics must suffice. Generally, the catalogues issued by specialized companies, which contain full particulars of components on the market, together with the range of sizes available, are a fruitful source of information.

The rational system of units and symbols (ISO), as defined in Volume 1, Chapter 1, has been used throughout.

Chapter 1

Structural Elements

A. PLAT

1.1 Explanatory note

This chapter does not go very deeply into the theoretical aspects of design, which are discussed in Volume 1, Chapter 2. Instead, the design results are demonstrated and discussed exclusively with reference to structural elements that are common to every precision engineering product.

Structural elements are, of course, the outcome of design work based on a specification that takes into account production capability and the cost.

Where necessary, mention is made of the engineering processes involved in the production of these parts: the cost is mentioned only in passing. Frequent references to technical and price considerations are to be found elsewhere in the Handbook.

The elements discussed are of a general nature. Individual subjects, such as plugs, switches, coils, transformers, relays and printed wiring are not mentioned in this chapter: particulars are given in Volume 7, Chapter 1. Specific equipment and instruments used in aircraft are also ignored here.

Diagrams are used wherever possible. They are vivid and have more impact on the designer than a full description in words. Since space does not permit exhaustive treatment of the subjects dealt with in this Handbook, only a few of the more important applications have been selected for discussion.

1.2 Permanent joints[1–14, 47–50]

1.2.1 Welded joints[15–28, 57–59]

Whenever a permanent joint is required, it is a good plan to consider whether a satisfactory result can be obtained by welding, since welded joints are fairly cheap to produce. The welding technique chosen depends on the material (e.g. metal or plastic) to be welded, whilst the type of joint depends on the structure of the product. Hence there is thus a wide range of possibilities, which can be classified according to the material, thickness and shape of the members to be joined. Actual welding processes are described in Volume 5, Chapter 2.

Metal parts can be welded together in many different ways. Resistance welding is preferred for joining sheet material, because it is fast and therefore economical. It is also used to join parts not made of sheet material. Joints made by ultrasonic welding are used to a limited extent.

Thermoplastics, in the form of foil, are usually welded by the high-frequency method or by applying heat (surges) from an outside source, when the dielectric losses involved are small.

Thicker parts are joined by means of a heating element, with or without a filler. Ultrasonic welding can also be used for this purpose. Solids of revolution made of thermoplastics can be joined by friction welding. See also Volume 3, Chapter 1.

Welds suitable for joining metal parts of various shapes and variously-shaped thermoplastic parts will now be shown one after another.

The information on the welds is clarified in some cases by particulars of the shaping of the joint members.

The captions to the illustrations are self-explanatory. Examples are given of welded joints between metal parts (Figs. 1.1 to 1.10 inclusive), and between

Fig. 1.1. Welding metal foil to solid metal parts.

    a. Foil to foil: percussion welding and ultrasonic welding.
    b. Foil to sheet: ultrasonic welding.
    c. Foil to sheet: spot welding, percussion welding or ultrasonic welding.
    d. Foil to rod: percussion welding or ultrasonic welding.

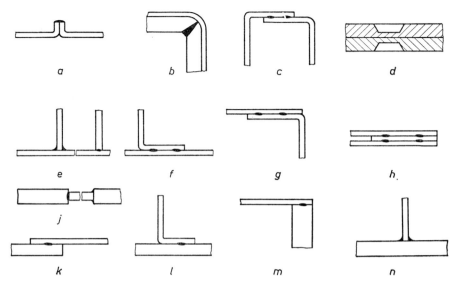

Fig. 1.2. Welding thin sheet metal.

a. Thin sheet to thin sheet: arc welding.
b. Thin sheet to thin sheet: arc welding.
c. Thin sheet to thin sheet: spot welding, projection welding or seam welding.
d. Thin sheet to thin sheet: cold pressure welding.
e. Thin sheet to thin sheet: arc welding or spot welding.
f. Thin sheet to thin sheet: spot welding or projection welding.
g. Thin sheet to thin angle bar: spot welding, projection welding or seam welding.
h. Three members of thin sheet material: spot welding or projection welding.
j. Thin sheet to thick sheet: arc welding or fusion welding.
k. Thin sheet to thick sheet: spot welding or projection welding.
l. Thin angle bar to thick sheet: spot welding, projection welding or seam welding.
m. Thin sheet to thick sheet: projection welding.
n. Thin sheet to thick sheet: arc welding.

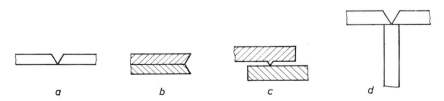

Fig. 1.3. Welding thick sheet metal.

a. Thick sheet to thick sheet: arc welding.
b. Thick sheet to thick sheet: arc welding.
c. Thick sheet to thick sheet: projection welding.
d. Three members made of thick sheet material: arc welding.

Fig. 1.4. Welding two sections together: cold pressure welding.

a   *b*

Fig. 1.5. Rough-machined metal fixing parts.
a. Bolt for projection welding.
b. Weld nut for projection welding.

parts made of thermoplastics (Figs. 1.11 to 1.14 inclusive). Methods of welding thermoplastics are shown diagrammatically in Figs. 1.15 to 1.19 inclusive.

1.2.2 Soldered joints[30-38, 57]

A soldered joint is permanent, but, unlike most other joints in this category, it can in many cases be detached, although with some difficulty. Soldering is classified according to the technique employed (Volume 5, Chapter 3.7), the various methods producing a variety of structures suited, as far as possible, to the particular purpose (Volume 5, Chapter 3.5). The suitability of different metals, alloys and coatings for soldering is discussed in Volume 5, Chapter 3.6. Care should be taken to position the members correctly when soldering them together. This can be done in various ways, as illustrated in Figs. 1.20 to 1.22 inclusive. An important element in soldering is the gap between the members. This gap must not be too wide, and its walls must be parallel, so as to form a capillary into which the solder can creep. Constriction of the gap in the direction of solder flow helps matters, but the gap must not expand in this direction (Figs. 1.23 to 1.26 and Figs. 1.41 to 1.43 inclusive). The area of solder must be in keeping with the structural strength required (Figs. 1.27 to 1.31 inclusive). When the load is fairly heavy, it is necessary to relieve the strain on the soldered joint (Fig. 1.32), which then serves simply as an attachment. Soldering is much used for electrical connections (Fig. 1.33) and solder tags of various shapes exist for this purpose (Fig. 1.34).

In the process of soldering, the work surfaces must be brought to the correct temperature, without in any way disturbing the surroundings. The joint members can be given a certain amount of thermal resistance, for example, by means of a suitable constriction or a blanked hole, to prevent the heat from draining away from the work surfaces, and, at the same time, to ensure that the ambient temperature is not raised unduly by the escaping heat (Fig. 1.35). A still more effective remedy is to employ a

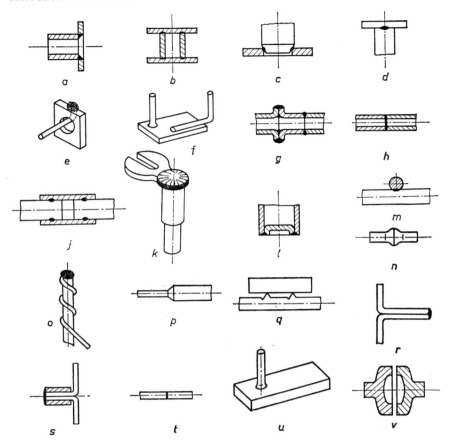

Fig. 1.6. Welding round metal parts.

a. Tube to sheet: arc welding.
b. Tube between two sheets: spot welding.
c. Pin through plate: fusion welding.
d. Pin to sheet: spot welding or projection welding.
e. Wire to strip: arc welding.
f. End of wire to face of sheet: percussion welding; side of wire to face of sheet: ultrasonic welding.
g. Pipe to pipe: arc welding.
h. Pipe to pipe: friction welding.
j. Rod in sleeve: spot welding.
k. Wire to cable tag: percussion welding or arc welding.
l. Bottom in drum: arc welding.

m. Rod to rod at right-angles: spot welding.
n. Two rods abutted: butt welding with pressure.
o. Wire to rod: arc welding.
p. Wire and bar abutted: percussion welding.
q. Disc to sheet: projection welding.
r. Wire to wire: arc welding, spot welding or percussion welding.
s. Wire to wire with sleeve: percussion welding or arc welding.
t. Two wires end-to-end: fusion welding.
u. End of wire to face of block: percussion welding.
v. Flange to flange: friction welding.

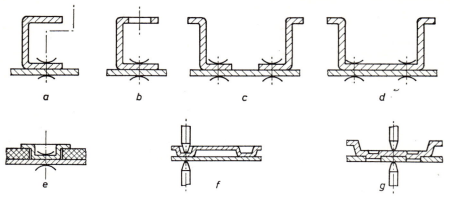

Fig. 1.7. Some combinations of parts to be welded together.

a. Spot welding with offset electrode.
b. Spot welding with electrode through hole.
c. Spot welding two Z-sections to sheet.
d. One channel is easier to spot weld than two Z-sections.
e. Spot welded cap secures insulating plate.
f. Embossed projections ensure a sound joint in spot welding.
g. Positioning by means of two centring bosses in two holes is often
employed in spot welding.

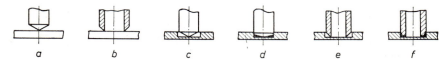

Fig. 1.8. Rods and pipes shaped for pressure butt-welding to sheet.

a. The tapered part of the rod functions as a projection.
b. The bevel on the pipe functions as a projection.
c. As *a*, but with a recess in the sheet.
d. As *c*, but after welding: the weld has a better appearance.
e. As *b*, but with a recess in the sheet.
f. As *e*, but after welding: the weld has a better appearance.

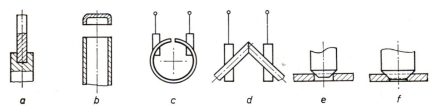

Fig. 1.9. Parts shaped for fusion welding.

a. Pin to bolt.   *b.* Cap on pipe.
c. Welding a ring by means of suitable electrodes.
d. Welding two rods at right-angles by means of suitable electrodes.
e. Pin to sheet, with hole for accurate location.
f. As *e*, after welding.

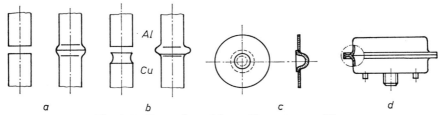

Fig. 1.10. Parts shaped for cold pressure welding.
a. Two aluminium rods before and after being welded.
b. An aluminium rod and a copper rod before and after being welded;
  the copper rod is grooved to flow better under pressure.
c. Cap and diaphragm, both made of foil.
d. Transistor capsule formed by two caps welded together.

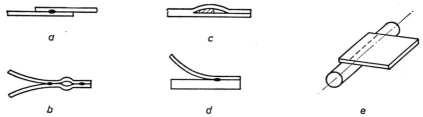

Fig. 1.11. Welding plastic foil to parts made of thermoplastic material.
a and *b*. Foil to foil: dielectric high-frequency welding, or welding with heat (surge).
c. Foil to foil: with heating element, without filler material.
d. Foil to sheet: dielectric high-frequency welding.
e. Foil to rod: dielectric high-frequency welding.

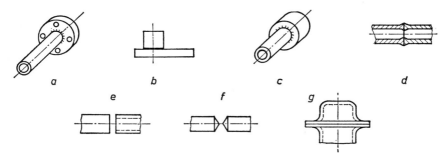

Fig. 1.12. Welding round parts made of thermoplastic material.
a. Flange to pipe: welding with heating element and filler material, or friction
  welding.
b. Stud on plate: welding with heating element without filler material, or friction
  welding.
c. Pipe to pipe: welding with heating element, without filler material.
d. Pipe to pipe: welding with heating element, without filler material, or friction
  welding.
e. Rod to pipe: friction welding.   *f*. Rod to rod: friction welding.
g. Cap to pipe: dielectric high-frequency welding.

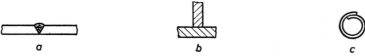

Fig. 1.13. Welding thermoplastic sheet material.

a. Sheet to sheet: welding with heating element or welding with hot gas, using filler material.

b. Sheet to sheet: welding with heating element, without filler material.

c. Tube formed of sheet material: dielectric high-frequency welding.

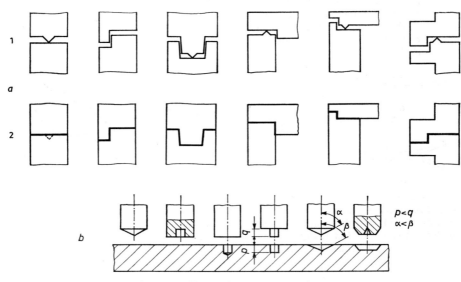

Fig. 1.14. Shapes for thermoplastic parts.

a. The ultrasonic welding of sheet: 1 = before welding; 2 = after welding.

b. Shapes employed for friction welding pins to sheet material.

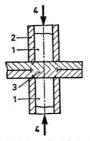

Fig. 1.15. Dielectric high-frequency welding.

1 = electrodes;
2 = pressure tool;
3 = melting zone;
4 = pressure applied.

Fig. 1.16. Seam welding two pieces of foil, whereby the surplus edge is trimmed off during welding.

1 = blade;
2 = shaping electrodes;
3 = scrap material.

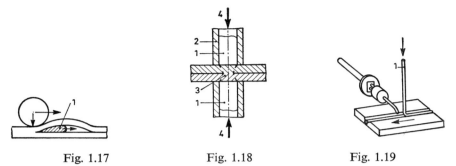

Fig. 1.17   Fig. 1.18   Fig. 1.19

Fig. 1.17. Welding with wedge-shaped heating element and pressure roller.
    1 = wedge-shaped heating element.

Fig. 1.18. Welding with heat surge.
    1 = electrodes; 2 = pressure tool; 3 = melting zone; pressure applied.

Fig. 1.19. Welding with hot gas and filler material.
    1 = supply of filler material.

heat-sink beyond the zone of thermal resistance, to protect the surrounds. A heat screen may also be effective in this respect (Fig. 1.36).

Allowance should be made for a possible difference in the coefficients of expansion of the joint members, which necessitates a somewhat thicker seam of solder. Differences in expansion of very small parts may safely be compensated (Fig. 1.37). Only compressive stresses are permissible in ceramic parts.

It is obvious that the designer can take full profit from the optimal construction of the soldering joints. A special example is controlling the shape of the compound component in Fig. 1.38, after cold pressing. The captions to the construction examples need no further explanation (Figs. 1.20 to 1.38 inclusive for soft soldering joints, and Figs. 1.39 to 1.43 inclusive for brazed joints).

1.2.3 Adhesive-bonded and cemented joints[39−46]

The physico-chemical aspects of such joints are discussed in Volume 2, Chapter 6 and the technology in Volume 5, Chapter 4.

These methods of joining are employed where others prove impracticable or not too satisfactory, for example where:

- joints made by other methods take up too much room or add unduly to the weight of the structure.
- a very uniform stress distribution is required in the joint members.
- other methods adversely affect the yield point, hardness, surface condition, electrical or magnetic properties.
- damping, elasticity or resistance to corrosion have to meet special requirements.
- a thin, decorative veneer or facing has to be applied.

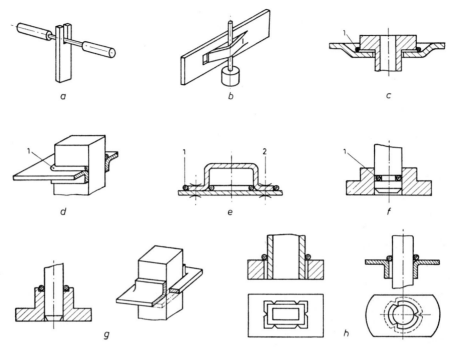

Fig. 1.20. Correct alignment of two parts to be soldered together.

a. Alignment by suitable shaping.
b. The parts hold each other in position.
c. Alignment of flange, hub and ring of solder.
d. Alignment of flange, shaft and ring of solder.
e. Alignment by spot welding.
f. Pin pressed into hub; ring of solder enclosed in groove.
g. The interference fit requires more expensive preliminary machining and assembly than the structure with spring tags.
h. Centring ridges also give parallel sides to the soldering gap.

1 = ring of solder; 2 = spot weld.

Fig. 1.21. Flux and solder applied by hand.

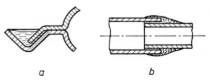

Fig. 1.22

a. Liquid solder has been poured into the trough.
b. Plastic solder has been wiped on.

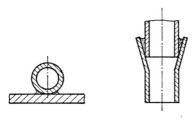

Fig. 1.23. Two examples of the
variation in gap widths.

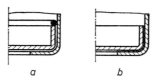

Fig. 1.24
a. Before furnace brazing.
b. After furnace brazing.

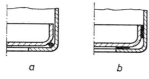

Fig. 1.25

a. Before furnace brazing: ring of
solder cannot escape.
b. Afterwards: solder has penetrated
into the narrow gaps.

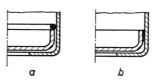

Fig. 1.26

a. Before furnace brazing.
b. Afterwards: the solder did not flow
from the narrow gap to the wide
one.

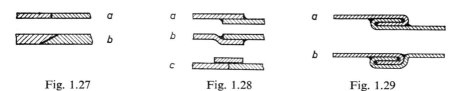

Fig. 1.27   Fig. 1.28   Fig. 1.29

Fig. 1.27. *a.* Butt soldered joint: relatively weak. *b.* Slightly stronger joint.

Fig. 1.28. *a.* Lap produces stronger soldered joint.
     b. Better than *a*, since walls joined are in-line (flush).
     c. Strong joint with strap.

Fig. 1.29. Soldering folded edges.
     a. Walls not in line. *b.* Walls in line.

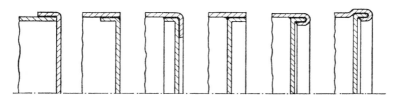

Fig. 1.30. Soldered flange joints for base and casing.

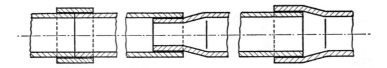

Fig. 1.31. Brazed joints for pipes.

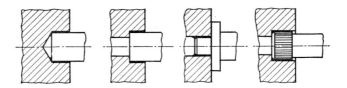

Fig. 1.32. Joints for brazing pin to wall: the joint is loaded.

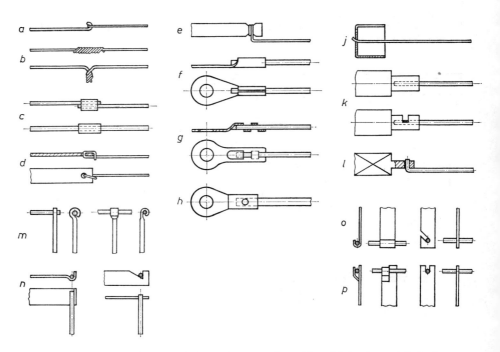

Fig. 1.33. Joints for electrical conductors: the soldered joints *a* to *p* inclusive
are relieved mechanically.

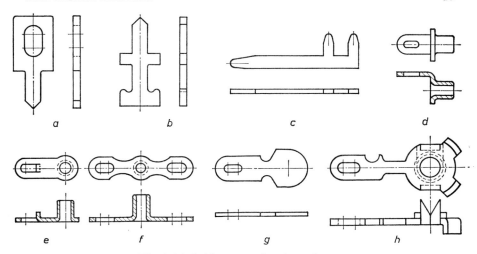

Fig. 1.34. Solder tags of various shapes.

a. For printed-wiring: on component side.
b. For printed-wiring: on print side.
c. For connecting small printed-wiring board to larger one.
d. Perpendicular solder tag.
e. Flat solder tag with lug to prevent twisting.
f. Double solder tag.
g. Solder tag suitable for spot welding.
h. Solder tag to which a connection can be screwed.

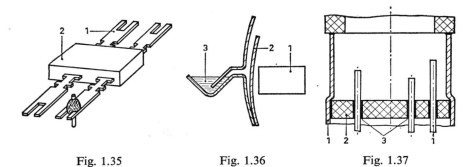

Fig. 1.35   Fig. 1.36   Fig. 1.37

Fig. 1.35. Solder tags with constrictions to prevent heat from draining
away from the soldering point.
1 = solder tag; 2 = thermoplastic.

Fig. 1.36. Heat screen 2 protects 1 from excessive heat.
3 = solder.

Fig. 1.37. Soldered joints suitable for ceramic parts.
1 = metal; 2 = ceramic; 3 = solder.

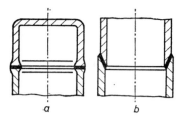

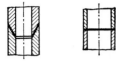

Fig. 1.38. Soldered joint obtained by cold-pressing.

Fig. 1.39. Brazed joints for pipes in line (abutted).

a. Size and location of bulge are not controlled.

b. Suitable shaping prevents bulging, so that joint takes up less room.

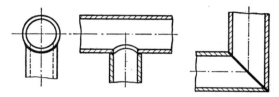

Fig. 1.40. Brazed joints for pipes at right-angles.

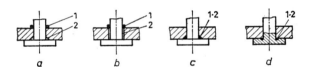

Fig. 1.41. Joint for brazing journal to wall.

a. Wrong: narrow solder gap must not be interrupted by cavity.
b. Correct: wide gap tapers into narrow one.
c. Correct: narrow gap in both directions of flow.
d. Correct: narrow gap in both directions of flow.

1 = ring of solder; 2 = cavity.

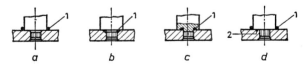

Fig. 1.42. Brazing knurled journal in hole in wall.

a. Wrong: solder can only reach knurl through narrow gap.
b. Correct: solder flows in both directions.
c. Correct: solder flows in both directions.
d. Correct: shoulder has radial knurl merging into finer (straight) knurl.

1 = ring of solder; 2 = radial knurl.

Fig. 1.43. Brazing fully-threaded and partially-threaded studs to wall.

  a. Wrong: solder also creeps up along thread.
  b. Correct: solder does not creep up.
  c. Correct: solder does not creep up.
  1 = ring of solder.

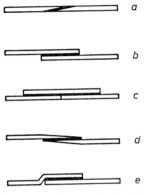

Fig. 1.44. Shaping adhesive-bonded joints to transmit shear stresses.

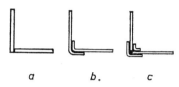

Fig. 1.45. Adhesive-bonded corner joints.

  a. Not recommended: adhesive covers too small an area.
  b. Recommended: adhesive covers a sufficient area.
  c. Still better: readily absorbs forces in both directions.

Fig. 1.46. Possible ways of preventing adhesive-bonded strip from peeling off.

  a. Possibility of peeling.
  b. Prevented by solid or tubular reinforcing rivet.
  c. End folded round edge.
  d. Area of adhesive enlarged at end of strip.
  e. Extra strap bonded on gives added rigidity.

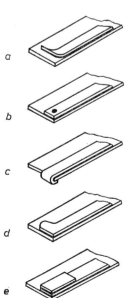

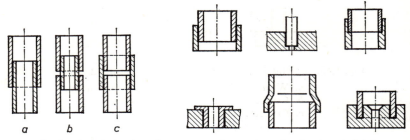

Fig. 1.47. Adhesive-bonded joints for pipes.

Fig. 1.48. Adhesive-bonded joints between variously shaped parts.

Fig. 1.49   Fig. 1.50   Fig. 1.51   Fig. 1.52

Fig. 1.49. An adhesive-bonded joint combined with a spot weld sets without being clamped; the load distribution is unfavourable because of the rigidity of the spot weld.

Fig. 1.50. An adhesive-bonded joint combined with a rivet sets without being clamped; the load distribution is unfavourable because of the rigidity of the rivet.

Fig. 1.51. An adhesive-bonded joint combined with an elastic grommet sets without being clamped, and is flexible.

Fig. 1.52. Adhesive-bonded seam (leakproof).

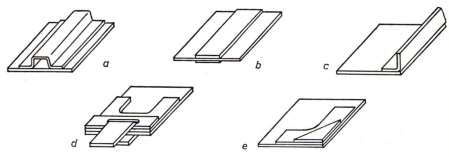

Fig. 1.53. Joints stiffened by additional members bonded on.

a. By hat section.
b. By two straps, to limit overall depth of joint.
c. By angle along edge.

d. Edge reinforced to withstand local loading.

e. Corner reinforcement.

Use such joints functionally, but try to keep down the cost of the process chosen by shaping the joint members to the best advantage, particularly so that the adhesive can be applied easily and does not take very long to set, and so that the surplus is not a problem and the joint can be made without expensive jigs, etc. Better temperature resistance of the joint members allows hardening at higher temperatures. Some examples of adhesive-bonded joints appear in Figs. 1.44 to 1.58 inclusive, with explanatory captions. One or two cemented joints are illustrated in Figs. 1.59 to 1.61 inclusive.

Fig. 1.54. Wooden joints: members in line.

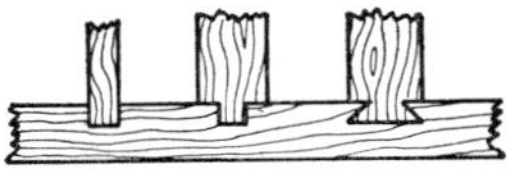

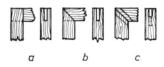

Fig. 1.55. Wooden joints: T-shaped.

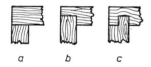

a   b   c

Fig. 1.56. Wooden corner joints.
a. For light loads.
b. For loads in one direction.
c. For loads in both directions.

a   b   c

Fig. 1.57. Wooden corner joints for frames.
a. Simple structure.
b. Visible from one side only.
c. Tongue-and-groove gives added strength.

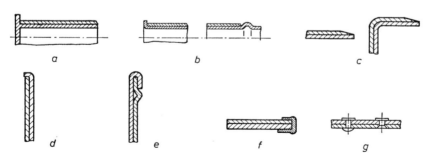

Fig. 1.58. Structures designed to prevent failure of adhesive-bonded joints.
a. Adhesive-bonded layer in optical instruments: flange prevents adhesive failure.
b. Layers bonded on to thin sheet: ridge prevents adhesive failure.
c. Thick, bonded layer, for example on leather: chamfered.
d. Flanged layer bonded to thin sheet.
e. Layer folded and bonded to thin sheet.
f. Layer of felt or thick fabric bonded on: edge finished off with frame.
g. Adhesive-bonded joint with decorative rivets.

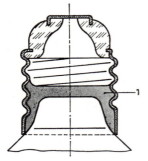

Fig. 1.59. Glass bulb cemented into screw cap of filament lamp.

1 = cement.

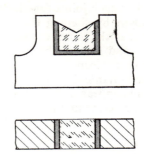

Fig. 1.60. Knife-edge bearing cemented in.

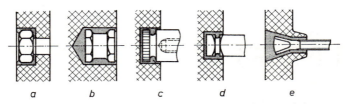

Fig. 1.61. Metal parts cemented in ceramic material.

a. Hexagonal nut.   *c.* Threaded pin.   *e.* End of wire.
b. Collar nut.   *d.* Square shaft.

1.2.4 Methods of joining by plastic deformation of material[51−54]

Particulars of the different techniques are given in Volume 5, Chapter 1. Methods of joining by plastic deformation of material are: riveting, seaming (flanging, curling and flattening), bending, folding, twisting, bead-forming or crimping, expanding or clinching, staking or indentation, stranding and wrapping.

Such joints are used, particularly in large-scale batch production, where the joint can be made in one operation. Whenever possible, extra fixing parts

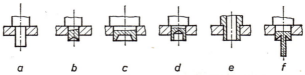

Fig. 1.62. Part of shaft deformed in fixing it in wall.

a. Solid shank for riveting to withstand heavy loads.
b. Recessed shank for lighter, axial loads.
c. As *b*, for larger diameter.
d. As *b* and *c*, with rim for riveting over.
e. With rim for riveting over, for lighter loads.
f. Form used for through shaft.

or fasteners are dispensed with. This is particularly important for machine-made joints. With a few exceptions, the joints are permanent.

(a) *Joints made by deforming part of a member shaft*

Figures 1.62 and 1.63 illustrate possible ways of shaping the member to be deformed, in structures that do not involve locking to prevent rotation. Structures whose members are locked to prevent them from rotating relative to one another are shown in Fig. 1.64 (rectangular members), Fig. 1.65 (turned members), Fig. 1.66 (specially-shaped holes) and Fig. 1.67 (special structures).

The further the locking action is away from the centre of rotation, the more effective is it in preventing rotation.

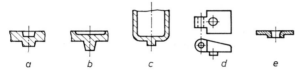

Fig. 1.63. Special parts for fixing to a wall with:

a. Coined boss.   c. Extruded lug.   e. Dimpled hole in wall.
b. Pressed lug.   d. Cast lug.

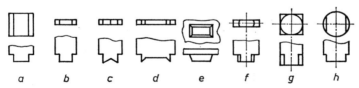

Fig. 1.64. Various angular-shaped bars and shafts for attachment by riveting, prevented from rotating by:

a. Milled end.   f. Turned spigot on strip material.
b, c and d. Stamped ends.   g. Turned journal on square material.
e. Pressed or cast boss.   h. Flats milled on round shaft.

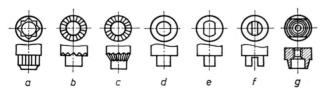

Fig. 1.65. Round shafts for attachment by riveting, prevented from rotating by:

a. Straight knurl.   d. Milled flat.   f. Milled slot.
b. Radial knurl.   e. Two milled flats.   g. Flats milled on nut.
c. Conical knurl.

(b) *Joints made by deforming an extra fixing part*

Rivets used for this purpose are almost invariably round, and of such relatively soft material as: mild steel, brass, copper, pure aluminium or aluminium alloy. Some of the solid rivets commonly used are shown in Fig. 1.68

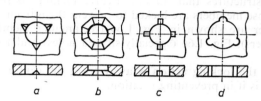

Fig. 1.66. Shafts for attachment by riveting, prevented from rotating by special holes in wall:

    a. Three notches in circumference of hole.
    b. Octagonal countersunk recess in hole.
    c. Four slots in circumference of hole.
    d. Hole punched over three smaller holes pierced beforehand.

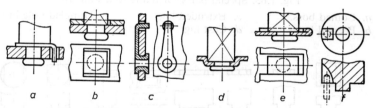

Fig. 1.67. Special structures to prevent rotation.

    a. Bracket, spot welded to shaft, whose flange drops into a hole.
    b. Head cup, spot welded to wall.
    c. Bush with flanged stem.
    d. Square shaft drops into matching depression in wall.
    e. Rectangular shaft drops into slot in part of wall folded back.
    f. Pin fitted in wall engages in slot in shaft.

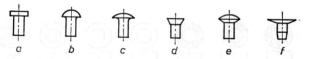

Fig. 1.68. Solid rivets.

    a. Flat-headed rivet: for simple riveting.
    b. Button-head rivet: for steel structures.
    c. Truss-head rivet.
    d. Rivet with countersunk head: hole for this is expensive to make.
    e. Rivet with raised (round top) countersunk head: good appearance.
    f. Rivet with flat countersunk head: for belts, etc.

Figure 1.69 shows various hollow rivets and Fig. 1.70 a number of special rivets. The choice of rivet depends very much on the strength, appearance and degree of accessibility required. One or two staples for use in softer materials appear in Fig. 1.71. Figs. 1.72 and 1.73 illustrate the use of rivets with countersunk driven heads.

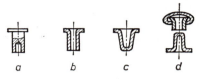

Fig. 1.69. Hollow rivets.

a. Semi-tubular rivet with flat head: for simple work.
b. Tubular rivet.
c. Drawn, closed rivet used, say, for lightproof and dustproof structures.
d. Two-part rivet: for very soft material.

Fig. 1.70. Special rivets.

a. Mushroom-head rivet: see also Fig. 1.73*b.*
b. Split, or bifurcated, rivet: the prongs can be forced apart.
c. Explosive rivet: see also Fig. 1.94*d.*
d. Rivet grips when pin is driven in: for structures accessible from one side only; also available in plastics.
e. Used as *d*: see also Fig. 1.94*c.*
f. Used as *d*: see also Fig. 1.94*a.*
g. Eyelet or fixing bush.

Fig. 1.71. Staples for fixing soft materials.

a. Strong round-wire staple.   *b* and *c.* Staples made of strip material.

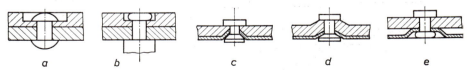

Fig. 1.72. Countersunk rivet heads.

a and *b.* Driven heads in countersunk holes.
c, d and *e.* Joints for fastening thin sheet to thick.

Fig. 1.73. Countersunk heads in thin sheet.

a. For soft material: light loads.

b. Sheets are indented during riveting of driven head.

c. Thin sheet shaped to give good appearance.

(c) *Indented joints*

The indentations can be obtained by centre-punching, scoring, etc., usually in only one of the joint members.

Variations of the method as applied to tubular, sheet, strip and bar material are shown in Fig. 1.74, where both members are indented; in Fig. 1.75, where a ridge or bead formed on one part drops into a slot in

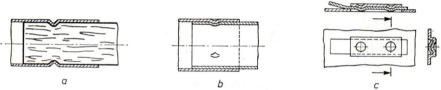

Fig. 1.74. Joints involving indentation of both members.

a. Wooden rod fixed in pipe by centre-punching.

b. Two pipes mated and fixed by scoring.

c. Strip fastened to sheet by centre-punching in two places.

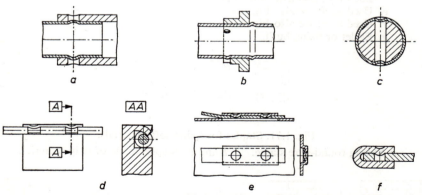

Fig. 1.75. Indentations in one member drop into slots in the other.

a. Thin-walled pipe in thick-walled pipe.

b. Pipe in flange.

c. Rod fixed in pipe by centre-punching.

d. Shaft fixed to plate by two indentations.

e. Strip fixed to plate by centre-punching in two places.

f. Rim fastened along edge of plate by centre-punching.

the other; and in Fig. 1.76, where one structural member is completely embraced or retained by the other, without being deformed itself.

Much less force is required to stake the ends of bars than to rivet them. Staked joints, as illustrated for round shafts in Fig. 1.77 and for flat bars in Fig. 1.78, can withstand only relatively light axial-loads.

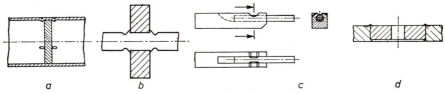

Fig. 1.76. Parts gripped by staking.
    a. Disc locked in pipe by centre-punching.
    b. Disc locked on shaft by staking.
    c. Pin in rod.
    d. Ring enclosed in wall by staking around periphery.

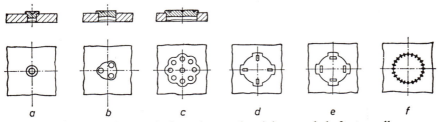

Fig. 1.77. Various staked notches used to join round shafts to walls.
a. Central notch for $d < 2$ mm.
b. Three centre-punched notches for $d > 2$ mm.
c. Several centre-punched notches to withstand heavier loads: the deforming forces involved are less than in riveting.
d. For large-diameter bars, the deforming forces required for radial notches are relatively small.
e. Tangential notches to transmit strong forces.
f. Numerous radial notches to limit the deforming force: only minor axial forces can be transmitted.

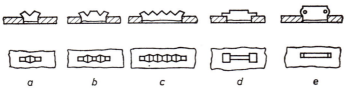

Fig. 1.78. Various notches used to join bars of rectangular cross-section to a wall.
    a, *b* and *c*. The number of notches depends on the width.
    d. Both sides staked: bar can pass on.
    e. Staked sideways: bar can pass on.

(d) *Seamed joints*

(For particulars of tools required, see Volume 5, Chapter 1.) Seam forming is done at projecting edges, usually integral with one of the joint members. Sometimes, the joint involves flanging and curling both members, as in Fig. 1.79. A number of examples are given: Fig. 1.80 shows thin sheet seams; Fig. 1.81 various seams used to join bottoms to cylinders. In Fig. 1.82, the flange is clenched or crimped into an indentation or slot in the other member, whilst Fig. 1.83 depicts various methods of forming stops for one member against the other. For brittle members, it is necessary to choose a construction that does not require much force to produce deformation. Pressure on a surface can be reduced by placing a suitable washer under the flange. Adhesive (Fig. 1.52) or solder (Figs. 1.29 and 1.30) must be applied to procure watertight or airtight seams.

(e) *Joints formed by bending or folding*

The parts bent or folded are usually integral with one of the joint members. Special constructional details ensure that there is no play in the joints (Figs. 1.84 to 1.88 inclusive). Other examples are given in Fig. 1.89, showing the fastening of tags to strips of insulating material; in Fig. 1.90, showing strips joined together; in Fig. 1.91, showing sheets joined together; and in Fig. 1.92, where a fastener is used to fix a rod in a pipe.

a   *b*

Fig. 1.79. These curled seams require expensive tools.

    a. Tube joined to sheet.
    b. Tube joined to joggled cover.

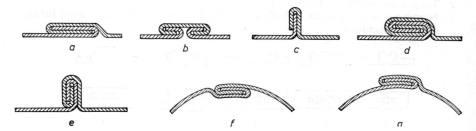

a   *b*   *c*   *d*

e   *f*   *a*

Fig. 1.80. Seams in thin sheet.

a. Longitudinal seam.	*d.* Longitudinal double seam.
b. Longitudinal seam with separate member.	*e.* Upright double seam.
	f. Inside seam.
c. Upright seam.	*g.* Outside seam.

The spring reflex of the tab bent over can be compensated by interposing some relatively soft material, such as pressboard, which also distributes the surface pressure more evenly. This is particularly important in the case of brittle parts.

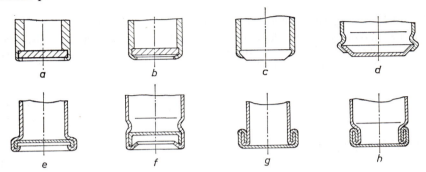

Fig. 1.81. Seams joining bottom to cylinder.

a. Ordinary flange formed by bending at right-angles.
b. Flange not at right-angles: cheaper but not so neat.
c. Bottom bevelled: usual for glass.
d. Crimped flange for a drawn base.
e. Seam for a drawn member: spinning the flange over 180° is expensive.
f. Similar to *e*, but cheaper.
g. Bottom folded together with edge of cylinder into tight seam.
h. As *g*, but cylinder afterwards expanded.

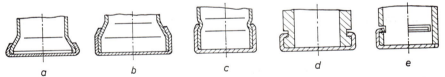

Fig. 1.82. Flanged joints, with flange of one member crimped into indentation or slot in the other.

    a. Flange encloses rim on cylinder.
    b. As *a*, but slightly more robust.
    c. Flange crimped into indentation in cylinder.
    d. Flange crimped into turned groove.
    e. Flange crimped into slots in pipe.

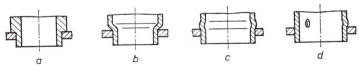

Fig. 1.83. Various ways of forming stops to hold pipe in hole in wall.

    a. Shoulder formed by turning recess in pipe.
    b. Constriction at end of pipe.
    c. Ridge formed on pipe.
    d. Pipe is notched or centre-punched.

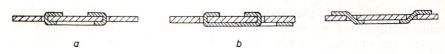

a

Fig. 1.84

b

Fig. 1.85

Fig. 1.84. Folded joint, without play: *a.* Staple. *b.* Solder tag.

Fig. 1.85. Folded joint that does not eliminate play completely, and in which the ends are flattened in the opposite direction in the second bending operation: not as strong.

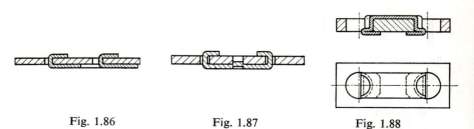

Fig. 1.86   Fig. 1.87   Fig. 1.88

Fig. 1.86. Play in this folded joint is unavoidable.

Fig. 1.87. Centring boss in hole eliminates play in joint.

Fig. 1.88. Play is eliminated from folded joint by subsequent clinching: fixing holes are round.

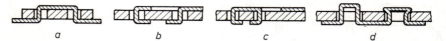

a   b   c   d

Fig. 1.89. Solder tags fastened to insulating strips.

a. Joint is not without play.
b and *c.* Play can be eliminated from these joints.
d. Multiple fastening.

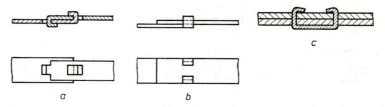

Fig. 1.90. Strips joined together by folding.

a. Strips are not exactly in-line.
b. Strips more nearly in-line: offset of lower strip may be less inconvenient.
c. Stapled joint for softer material.

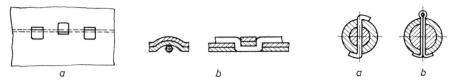

Fig. 1.91   Fig. 1.92

Fig. 1.91. Sheet material joined by bending.
   a. Sheets joined at right-angles by bent tabs.
   b. Sheets of soft material joined by wire bracket.

Fig. 1.92. Joining turned parts.
   a. With piece of bent wire.
   b. With split pin.

(f) *Twisted tab joints*

Because they are to some extent detachable, these joints are sometimes used to fasten members that may have to be replaced later, in order to repair the parent structure. But, they require a certain amount of vertical clearance above them. Fig. 1.93 gives one or two examples.

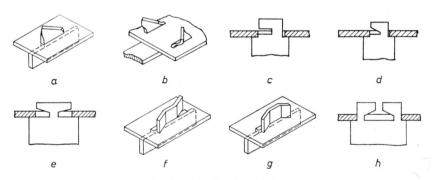

Fig. 1.93. Twisted tab joints.

a. Two strips at right-angles (T-joint).   e. As d: double tab.
b. Strip parallel to plate.   f. As e: tabs bent in opposite directions.
c. Single-sided tab.   g. As e: tabs bent in same direction.
d. As c: tab has tightening effect.   h. Better positioning.

(g) *Joints formed by expanding or beading*

They can be constructed in many different ways, and are used for structures accessible from one side only, as shown in Figs. 1.94 to 1.97 inclusive (blind assembly). Other examples are given in Figs. 1.98 to 1.100 inclusive. The joints illustrated in Figs. 1.101 to 1.103 inclusive are crimped.

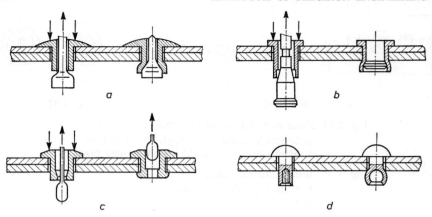

Fig. 1.94. Methods of expanding blind rivets, tubular rivets and eyelets
accessible from one side only.

a. Tool is drawn into rivet and then snapped off.
b. Tool is drawn into rivet: protruding end is then removed.
c. Tool is drawn right through hole.
d. Explosive rivet: the rivet, filled with explosive, is fired by heating it.

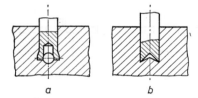

Fig. 1.95. Pin locked in hole by
expansion.

a. Expanded by ball: hole must be
soft-walled.
b. Expanded by cone at bottom of
hole: hole must be soft-walled.

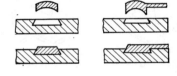

Fig. 1.96. Method of fixing a part in
a dovetailed slot.

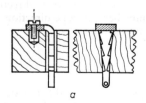

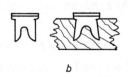

Fig. 1.97. Reinforcing members, e.g. for wooden wall.

a. Tapped hole fixture.   b. Staple, before and after being driven in.

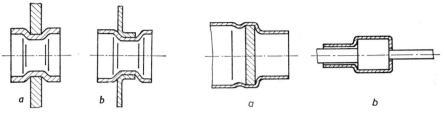

Fig. 1.98   Fig. 1.99

Fig. 1.98. Expansion joints.
    a. Pipe anchored in hole in wall.
    b. Pipe anchored more securely in wall, due to upright rim on hole.

Fig. 1.99. Crimped joints.
    a. Disc crimped in pipe.   *b.* Pipe crimped on to rod.

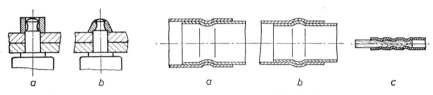

Fig. 1.100   Fig. 1.101

Fig. 1.100. Fastening parts by means of clenched retaining bush.
    a. Before clenching.
    b. After clenching: bush has been forced into groove in pin.

Fig. 1.101. Both joint members are beaded.
    a. Two pipes slotted together: bead inwards.
    b. Ditto: bead outwards.
    c. Rod of soft material crimped in a pipe.

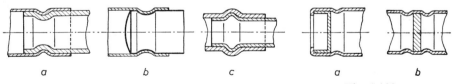

Fig. 1.102   Fig. 1.103

Fig. 1.102. Bead or rib of one member engages in preformed bead
    of other member.
    a. Pipe fixed round pipe by beads.
    b. Pipe fixed round rod by beads.
    c. Pipe locked in pipe by beads.

Fig. 1.103. Retention by beading.
    a. Cup in pipe.
    b. Partition retained in pipe by beads on both sides.

(h) *Wrapped joints*[57]

The wrap-around joint (Fig. 1.104) also serves as an electrical connection, because of the considerable contact force exerted on the sides of the prismatic rod by the round wire. Although it requires very careful construction, this joint is reliable and is used, amongst other things, in telephone exchanges. One or two examples appear in Fig. 1.105.

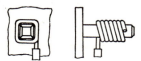

Fig. 1.104. Wrapped joint for electrical connections. Wire is wrapped round the square pin with great force.

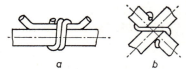

Fig. 1.105. Two wires joined by another wire.

a. Thin wire joined to a thicker wire.
b. Two crossed wires of equal thickness.

1.2.5 Interference joints[55, 56]

(See also Fig. 1.273 and Volume 5, Chapter 1.3.)

In joints of this kind, one member completely (or partially) encloses the other. The joint is sustained by the frictional force set up through elastic (and, in most cases, also plastic) deformation of the joint members, and cannot be unfastened without damaging them.

Similar joints can also be made with separate fasteners such as dowels, fluted pins, fluted rivets, spring dowels and wire nails.

The softer of the two members is deformed more than the other during forcing in; it also determines the strength of the joint. In addition, allowance must be made for possible shrinkage (for soft metals, thermoplastics) or ageing (for thermosetting plastics, diecastings).

The amount of interference depends on the size of the joint and the moduli of elasticity of the materials. For example, a brass member requires more interference than a steel member, and so does not have to be machined as accurately. Appreciably greater (and less exact) interferences can also be employed when one of the members is given a straight knurl. A bevelled lead-in, together with a knurl where necessary, should be cut on the harder member, usually a shaft or pin (Fig. 1.106). Special unknurled sections serve to centre the driven, knurled part more accurately (Fig. 1.107). After being forced in, the depressions between the high points of the knurl should not be completely filled (Fig. 1.108), or the forces set up will be too great. Examples of such joints are shown in Fig. 1.109 and, later, in Fig. 1.273.

Fluted pins, like knurled shafts, do not require such an accurate hole as cylindrical dowels. The more widely-used fluted pin has three full-length flutes. The flutes may also be tapered or may extend over only part of the length of the pin (Fig. 1.110). Examples of structures involving fluted pins are given in Figs. 1.111 and 1.112.

The resilience of a spring dowel enables it to be forced into holes still less accurate than those required for fluted pins (Fig. 1.113). A spring dowel is made of rolled, hardened and tempered sheet steel. It has many practical uses, for it does not shake loose and can be withdrawn and reinserted many times without losing its efficiency. Similar joints are made too with: screw nails, wire nails (for wood) and rivets (Fig. 1.114). For particulars of hole sizes for these fasteners, see the appropriate standards, which also specify the interference and the associated tolerance.

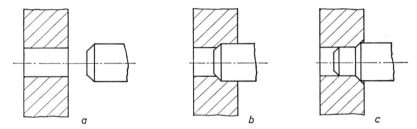

Fig. 1.106. Diagram illustrating process of driving one part into another.

a. Before driving in.
b. Because shaft is harder than hole wall, lead-in is made on it.
c. Hole wall is harder than shaft, and is therefore provided with lead-in.

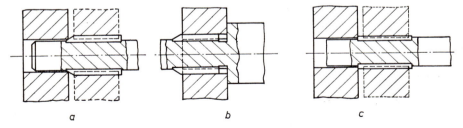

Fig. 1.107. Driving in the knurled end of a shaft.

a. Cylindrical centring shank leads knurl in driving direction. Dotted line represents position of wall after shaft has been driven in.
b. Cylindrical centring shank follows knurl in driving direction.
c. Long shaft with knurl at centre enables wall to slide smoothly over shaft and nevertheless grip tightly on the knurl. Dotted line represents position of wall after shaft has been driven in.

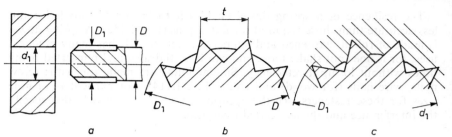

Fig. 1.108. Dimensions of members of interference joint.

a. Separate joint members.
b. Cross-section of knurled shaft (enlarged).
c. After driving in; section at right-angles to axis of joint. Driven with moderate force, groove between neighbouring high points of knurl is not completely filled.

Pitch and interference (mm).

D	t	$D_1 - d_1$
1·5–2·8	0·3	+ 0·06–0·12
3 – 4·5	0·5	+ 0·08–0·15
5 – 8	0·5	+ 0·12–0·20

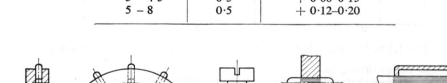

Fig. 1.109. Examples of structures involving interference joints.

a. Indexing wheel with dowels driven in.
b. Screw, with contact pin of (expensive) contact material driven into it.
c. Lever with partly knurled pin driven into it.
d. Knurled shaft with lever forced on to it.

1.2.6 Embedded inserts[56]

(See Chapters 5 and 6 of Volume 10 and Chapter 1 of Volume 3.)

Inserts are used in plastics and cast-metal products. Embedding usually takes place in metal moulds or impressions, because the insert often has to be positioned very accurately. Possible reasons for choosing this type of structure are:

● to impart some local or other property to the product, such as increased strength, hardness or resistance to wear;
● to procure solderability, resilience or thermal or electrical conductivity;
● to avoid complex cavities or cores in the mould.

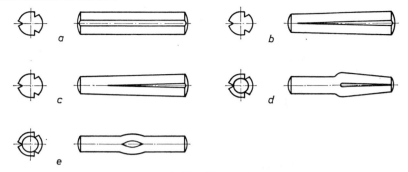

Fig. 1.110. Fluted pins.

a. Fluted pin with three full-length flutes.
b. As a, but flutes are tapered, so that pin acts as key.
c. As b, but flutes cover only half the length of the pin.
d. Fluted pin with three tapered flutes and a plain shank.
e. Fluted pin with three flutes at the centre and plain shanks at both ends.

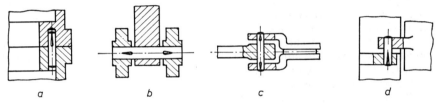

Fig. 1.111. Examples of structures involving fluted pins.

a. Fluted pin for fastening a detachable part.
b. Fluted pin for fastening two rollers.
c. Hinge with fluted pin.
d. Fluted pin as hinge pin.

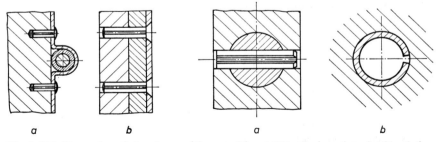

Fig. 1.112. Examples of structures with fluted rivets.

a. Bracket anchored with two round-head fluted rivets.
b. Fastening with countersunk fluted rivets.

Fig. 1.113. Spring dowel after being driven in

a. Side view.
b. Cross-section.

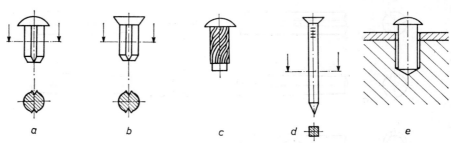

Fig. 1.114. Rivets and nails.

   a. Round-head fluted rivet (see also Fig. 1.112*a*).
   b. Fluted rivet with countersunk head (see also Fig. 1.112*b*).
   c. Drive screw or screw nail, for fixing type plates, etc.
   d. Wire nail, for wooden structures.
   e. Upsetting rivet is fixed in hole by upsetting: for type plates, etc.

The inserts, almost invariably of metal, are anchored in the moulding, injection moulding or casting compound or metal, by being given a shape that prohibits rotation or withdrawal. As a rule, there is no adhesive bond between the materials but, particularly in the case of diecast metal, the insert is anchored to some extent by shrinkage.

The wall thickness surrounding metal parts moulded or diecast in plastics must be sufficient to allow for the relatively greater shrinkage involved. Fig. 1.115 shows structures with nuts moulded into them, three of the inserts being ordinary hexagonal nuts and the other a cap nut.

Points to note are the location of the nut, possible fouling of the thread, and cost savings. Fig. 1.116 shows nuts which drop into the mould with a round projection, enabling them to be properly supported and deflashed. Fig. 1.117 depicts nuts accessible from both sides. The embedded parts in Fig. 1.118 are tapped afterwards to form nuts. Embedded studs are shown in Fig. 1.119. All these parts are anchored by knurls, hexagons, undercuts or flats. Fig. 1.120 details pins machined beforehand to anchor firmly. Blanked parts are easily anchored (Figs. 1.121 to 1.125 inclusive).

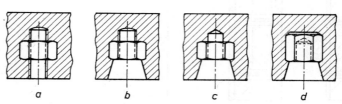

Fig. 1.115. Embedded hexagonal nut and cap nut.

   a. Nut screwed on to a threaded pin, which is a detachable part of the mould and must be unscrewed or backed-out of the nut after the moulding compound has hardened. Localization is inaccurate. The stud deforms readily.
   b. Nut rests on conical part: location is more precise and pin stronger.
   c. As *b*, but with nut fitted on a smooth, solid pin. Easier, but moulding compound may enter thread.
   d. Cap nut: compound does not enter thread.

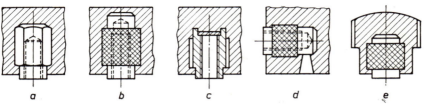

Fig. 1.116. Nuts suitable for embedding, with shoulder to provide support in mould and to facilitate deflashing.

a. Cap nut of hexagonal material.
b. Cap nut of round material, with diamond knurl.
c. Fully-threaded nut, sealed-off by disc pressed into recess at one end.
d. Axis of nut at right-angles to direction of moulding pressure, with extra support.
e. Shell of plastic must not be too thin, or it will crack.

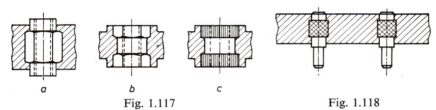

Fig. 1.117   Fig. 1.118

Fig. 1.117. Embedded nuts, accessible from both sides.
    a. Both ends for reference in mould.
    b. With shoulders, which are milled off after moulding.
    c. Thread is tapped after moulding; this enables bush to be placed in mould on a pin.

Fig. 1.118. Structure for exact spacing of threaded insert nuts. After moulding compound has hardened and shrunk, the pins used to hold inserts in mould are routed out, and thread is then tapped. If pin diameter is small enough, pins drop out during the pre-tapping drilling.

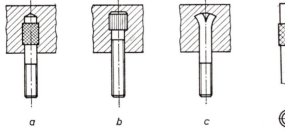

Fig. 1.119. Embedded studs.

a. Diamond knurl for anchoring.
b. Straight knurl: the moulding compound may enter the thread.
c. Flattened part for anchoring.

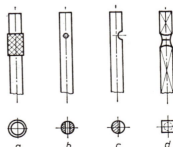

Fig. 1.120. Pins for embedding.

a. Round with diamond knurl.
b. Round with lateral hole.
c. Round with milled notch.
d. Square with turned groove.

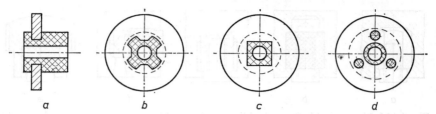

Fig. 1.121. Blanked disc moulded on to hub.

a. Longitudinal section.
b. Cruciform hole prevents disc from twisting on hub.

c. Ditto, with square hole.
d. Ditto, with three holes.

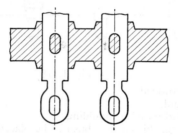

Fig. 1.122. Two blanked solder tags embedded; the creepage between them is enhanced by upright edges round the tags.

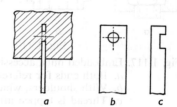

Fig. 1.123. Structures with embedded tag.

a. Longitudinal section.
b. Tag with hole for anchoring.
c. Tag with slot for anchoring.

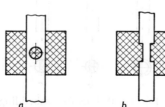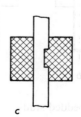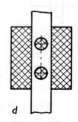

Fig. 1.124. Structures with double-ended tags extending through them, anchored by means of:

a. Round hole.   b. Two slots.   c. One slot.   d. Two holes.

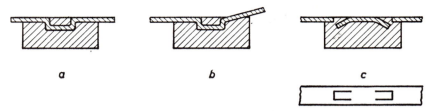

Fig. 1.125. Structure with bent part embedded in it.

a and *b*. With U-shaped bend.   *c*. With two separated tags.

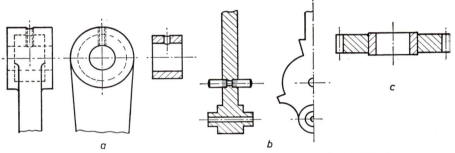

Fig. 1.126. Light metal diecast structures with parts embedded in them.

    a. Brass bush in crank.   *c*. Brass bush in gear wheel.
    b. Steel pin in switch lever.

The metals most commonly used in diecasting are zinc or aluminium alloys. Because of the relative strength and very low viscosity of these molten metals during diecasting, inserts anchor better in them than in many plastics. Fig. 1.126 shows one or two examples of such structures, whilst Fig. 1.127 illustrates the assembly of a disc on a shaft by embedding it in a light metal alloy.

Fig. 1.127. Rotary disc mounted on spindle for electricity supply meter. Shaft and disc are embedded simultaneously and thus united; pin is knurled with three holes in the disc.

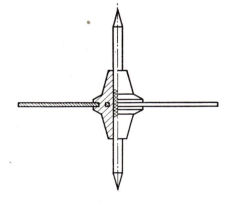

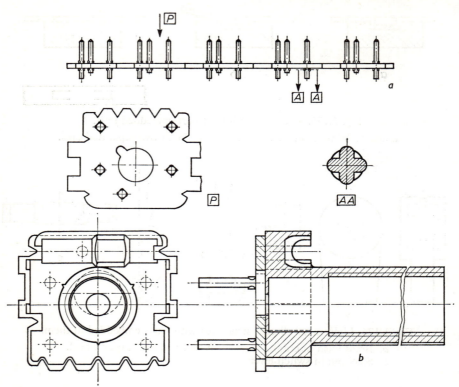

Fig. 1.128. Five pins assembled in a sheet of insulating material, moulded into the base of a coil former in a six-impression mould.

a. Group of six pin-and-plate assemblies ready for placing in the six-impression mould.

b. Complete coil former after being detached from the six-part moulding.

The insulating plate is for easy handling, and to reinforce the fastening of the pins in the base of the former.

The parts to be embedded have to be placed in a mould that is often hot, a difficult, time-consuming and responsible task. The problem can to some extent be overcome by combining several individual parts in a single unit beforehand, for example, by:

- Mounting individual electrical contact pins on a synthetic resin bonded cardboard plate, and placing this assembly in the mould (Fig. 1.128).
- Leaving blanked tags attached to the strip, moulding them in together and separating them afterwards (Fig. 1.129).
- Placing a circular arrangement of individual parts in the mould as a single unit and afterwards removing the segments uniting these parts (Fig. 1.130).
- Adopting an in-line arrangement of multiple impressions for moulded products with pins, so that the number of wires placed in the mould need only equal the number of pins. After moulding, the multiple product

can if necessary be processed so that the wires are severed and in some cases bent over (Figs. 1.131 and 1.132). The pins are anchored by knurling. A pin can be further secured against rotation by bending it and dropping the bent part into a slot.

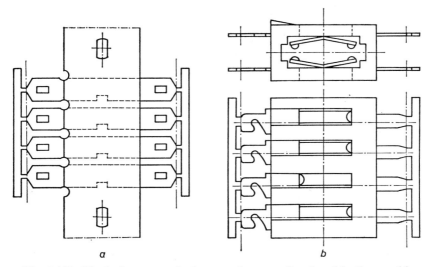

Fig. 1.129. Blanked tags attached to a strip are easily placed in the mould. After moulding, the tags are severed from the strip.

a. Solder tags on a bar.
b. Two rows of contact springs in stator of slide switch.

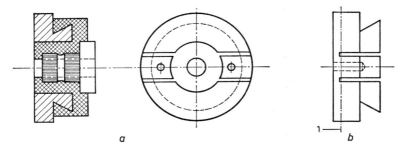

Fig. 1.130. Commutator with embedded parts.

a. Longitudinal section and front view.
b. Part to be embedded, with saw slots.

After hardening, the product is turned off as far as plane 1, thereby separating the four metal parts.

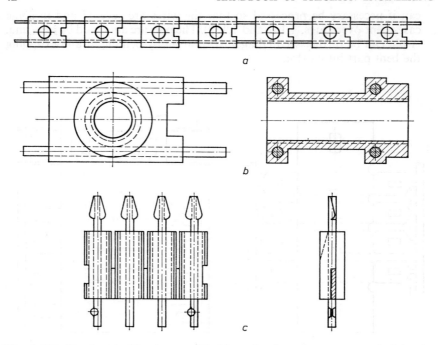

Fig. 1.131. Products with pins moulded-in, whereby wires are stretched in the (multiple-impression) mould. In this way, a multiple product is moulded, with the pins in line. After moulding, the products are separated and the projecting wire is processed if necessary.

  a. Multiple product after being moulded.
  b. Coil former with four pins.
  c. Contact strip with four processed pins.

Other methods of facilitating the insertion of pins in moulds for injection moulded products are:

● Closing the mould and then feeding the wire into the impression as far as a stop. The wire is trimmed to the required length after moulding (Fig. 1.133*a*) and, if necessary, bent over after the mould is opened (Fig. 1.133*b*), whereby (if possible) the bent part is dropped into a slot to prevent rotation. Knurls, preserving the round cross-section of the wire, ensure anchorage.
● Feeding prefabricated pins into the mould from a hopper. The pins may be given valvular heads for anchoring and to seal the mould (Fig. 1.133*c*).

If such an (economical) solution to the problem is not feasible, then embedding can sometimes be avoided altogether by fitting the additional part later, that is, by:

● Pressing it into, or riveting or clenching it on to, the moulding, etc.

- Screwing it in, if it is provided with an external self-tapping thread.
- Glueing, cementing or sealing.
- In the case of a nut, by insertion in a moulded slot shaped to prohibit rotation.

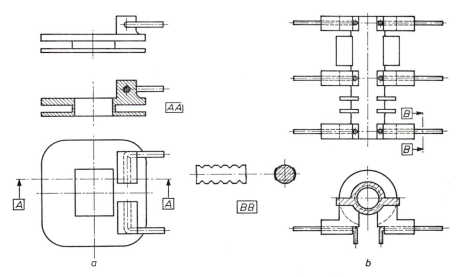

Fig. 1.132. Coil formers with bent pins moulded into them, whereby wires are clamped in the (multiple-impression) mould.

a. Coil former with two pins, moulded with the two pins straight and in line: after moulding, the pins are separated and bent over.

b. Coil former with six pins, moulded with pairs of straight pins in line: after moulding, the pins are separated and bent over.

Several parts can be assembled by placing them in a mould and partially investing them together (Fig. 1.127 and Figs. 1.134 to 1.136 inclusive). This assembly process also gives scope for the use of cost-saving techniques for positioning the individual parts in the mould or impression. Substantial savings can be achieved through such an assembly with (in many cases) plastics because they permit a considerable reduction in the number of components involved and the assembly costs, and, at the same time, facilitate mechanical assembly (see also Chapter 2 of Volume 8).

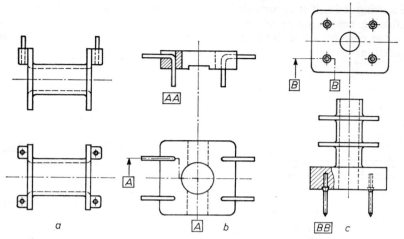

Fig. 1.133. Products with pins moulded into them, without combining the pins into a single unit to facilitate their insertion in the mould.

a. Coil former with four pins: during insertion, pins are taken from four spools of wire.

b. Coil former base with four pins; method as *a*, but, in addition, the four pins are bent over when the mould is opened.

c. Coil bobbin with four valvular pins fed into the mould from a hopper.

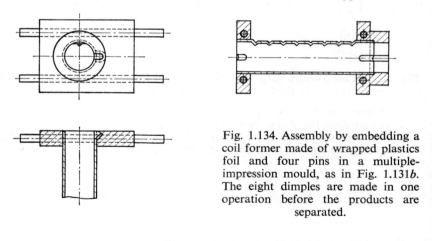

Fig. 1.134. Assembly by embedding a coil former made of wrapped plastics foil and four pins in a multiple-impression mould, as in Fig. 1.131*b*. The eight dimples are made in one operation before the products are separated.

Fig. 1.135. Assembly of a shaft and an annular magnet by embedding.

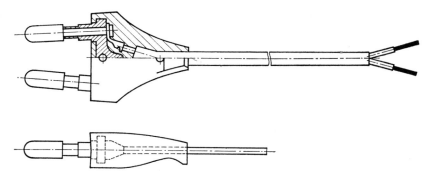

Fig. 1.136. Plug formed by embedding a cable together with plug pins.

REFERENCES

[1] RICHTER – V. VOSS und KOZER, *Bauelemente der Feinmechanik*, Verlag Technik, Berlin, 1954.

[2] K. HAIN, *Die Feinwerktechnik*, Fachbuch Verlag Dr. Pfanneberg, Giessen, 1953.

[3] G. SCHLEE, *Feinmechanische Bauteile*, Verlag Konrad Wittwer, Stuttgart, 1950.

[4] D. C. GREENWOOD, *Product Engineering Design Manual*, McGraw-Hill, New York/Toronto/London, 1959.

[5] F. WOLF, *Die nicht-lösbaren Verbindungen in der Feinwerktechnik*, Deutscher Fachzeitschriften- und Fachbuchverlag, Stuttgart, 1955.

[6] K. H. SIEKER, *Fertigungs- und Stoffgerechtes Gestalten in der Feinwerktechnik*, Springer Verlag, Berlin/Göttingen/Heidelberg, 1954.

[7] H. EDER and W. P. UHDEN, *Taschenbuch der Feinwerktechnik*, C. F. Winter'sche Verlaghandlung, Prien, 1965.

[8] K. RABE, *Grundlagen Feinmechanischer Konstruktionen*, Anton Zeimsen Verlag, Wittenberg/Lutherstadt, 1942.

[9] D. C. GREENWOOD, *Engineering Data for Product Design*, 1961.

[10] *Fastening and Joining*, Reference Issue, Machine Design, 1967, June 15.

[11] *The Fasteners Book*, Machine Design, 1960, September 29.

[12] R. PÖSCHL, *Verbindungselemente der Feinwerktechnik*, Springer Verlag, Berlin/Göttingen/Heidelberg, 1954.

[13] J. SOLED, *Fasteners Handbook*, Reinhold, New York/London, 1957.

[14] LAUGHER and HARGAN, *Fastening and Joining of Metal Parts*, McGraw-Hill, New York/London, 1956.

[15] F. EICHHORN and H. J. OPPE, *Das Verbindungsschweissen von Kleinstbauteilen*, Feinwerktechnik, 1967, 10, p. 449.

[16] H. EDER and W. P. UHDEN, *Feinwerktechnische Verbindungen durch Schweissen*, Feinwerktechnik, 1963, 4, p. 110.

[17] P. L. J. LEDER, *A Review of some Recent Developments in Welding Processes*, Journal of the Institute of Metals, 1964–1965, p. 375.

[18] D. S. PRESTON and T. K. HENDERSON, *Wire Wrapping Saves Time*, Product Engineering, 1961, November 13, p. 86.

[19] D. PRESTON, *Welded Connections*, Product Engineering, 1963, August 5, p. 78.

[20] W. A. OWCZARSKI, *Getting the most from Projection Welding*, Machinery, October 1962, p. 97.

[21] J. WODARA, *Widerstandsschweissen von Kleinteilen*, Schweisstechnik, 1962, January, p. 27.

[22] H. Böhme, *Die Widerstandsschweisstechnik in der Feingeräteindustrie*, Feingerätetechnik, 1963, H. 10, p. 456.

[23] W. Hofmann and F. Burat, *Fortschritte auf dem Gebiete der Kaltpressschweissung*, Feinwerktechnik, 1959, 3, p. 82.

[24] G. Maronna and B. Weiss, *Das Ultraschallschweissen-ein Überblick*, Schweisstechnik, 1965, H. 4, p. 167.

[25] M. B. Hollander, C. J. Cheng and J. A. Quimby, *Friction Welding*, Machine Design, June 18, 1964, p. 212.

[26] J. G. Steger, *La soudure ultrasonique des matières thermoplastiques*, Mécanique électricité, Février 1965, p. 57.

[27] J. King, *Which Way to Weld? Resistance, TIG or Electron Beam . . .*, Product Engineering, 1964, October 12, p. 77.

[28] L. Schade, *Die Auswirkungen eines Zinnüberzüges auf den Schweissprozess beim Widerstandsschweissen von dünnen stahlblechen*, Blech, 1963, p. 518.

[29] N. N., *Resistance: First Choice for Wide Range of Miniature Joining*, Welding Engineer, 1965, p. 42.

[30] H. Angermaier and H. Eder, *Feinwerktechnische Verbindungen durch Löten*, Feinwerktechnik, 1964, H. 4, p. 141.

[31] H. H. Manko, *How to Design the Soldered Electrical Connection*, Product Engineering, 1961, June 12.

[32] J. Colbus, *Verbindungen durch Löten*, Lastechniek, 1962, July, p. 149; September, p. 209.

[33] *Designing for Preforms*, Lucas-Milthaupt Engineering Co., Cudahy, Wisconsin.

[34] E. Löder, *Handbuch der Löttechnik*, Verlag Technik, Berlin, 1952.

[35] J. Schutz, *Die metallurgische Vorgänge zwischen Hartlot und Grundwerkstoffen und Folgerungen für die lötgerechte Konstruktion*, Schweissen und Schneiden, 1957, 12, p. 522.

[36] J. Colbus, *Gründsätzliche Fragen zum Löten und zu den Lötverbindungen*, Konstruktion, 1955, H. 11, p. 419.

[37] A. S. Cross Jr., *Brazing with Preforms*, Product Engineering, 1961, October 30, p. 45.

[38] E. P. Kuhlmann, *Hartlöten von Präzisionsteilen im Schutzgasofen*, Feinwerktechnik, 1955, H. 11, p. 391.

[39] H. Eder and W. P. Uhden, *Feinwerktechnische Verbindungen durch Kleben*, Feinwerktechnik, 1966, 11, p. 529; H. 12, p. 567.

[40] D. M. Weggemans, *Construeren met gelijmde verbindingen*, Plastica, 1963, p. 195 and p. 272.

[41] A. Matting and K. Ulmer, *Spannungsverteilung in Metall-Klebverbindungen*, VDI-Z, 1963, 13, p. 449.

[42] A. Matting and G. Hennig, *Das Metallkleben in der Einzel- und Mengenfertigung*, Feinwerktechnik, 1966, 1, p. 9.

[43] *Loctite technik*, VIBA, Den Haag.

[44] F. Mittrop, *Untersuchungen über die Kombination von Metallkleben und Punktschweissen*, Mitteilungen Forschungsgesellschaft Blechverarbeitung, 1965, 3.

[45] J. Eilers, *Kleben als Verbindungsverfahren für Kunststoffe*, Kunststoffe, 1961, 9, p. 611.

[46] K. F. Hahn, *Die Metallklebtechnik vom Standpunkt des Konstrukteurs*, Konstruktion, 1956, 4, p. 127.

[47] K. Meyerhans, *Das Verbinden von Metallen unter sich oder mit anderen Werkstoffen*, Metall, Sonderdruck 9/10, 1952, p. 229.

[48] *Fasteners for Packaging/Production*, Special Fastener Supplement of Electronic Packaging and Production, volume 6 section 2, 1966, September.

[49] M. W. RILEY, *Joining and Fastening Plastics*, Materials in Design Engineering 47, p. 129.

[50] W. A. GLEASON, *Five Ways to Seal Glass to Metal*, Materials in Design Engineering, 1960, April, p. 120.

[51] H. EDER, *Feinwerktechnische Verbindungen durch plastisches Verformen*, Feinwerktechnik, 1961, 4, p. 135.

[52] E. BÜRGER, *Nabenverbindungen im Gerätebau*, Feinwerktechnik, 1965, 12, p. 570.

[53] T. C. BUCHANAN, *Kleine Niete für Massenfertigung*, Konstruktion, 1957, 3, p. 126.

[54] W. C. MILLS, *Blechverbindungen*, Konstruktion 1957, 3, p. 124.

[55] J. KREISSIG, *Spritzgiessen in der Montage*, Werkstattstechnik 1965, 3, p. 139.

[56] Catalogue Prestinsert, Enfield, Middlesex, England.

[57] G. W. MILLS, *A Comparison of Permanent Electrical Connections*, Assembly and Fastener Methods, 1965, July, p. 48.

[58] *The Case for cold Welding*, Electronic Packaging and Production, 1969, March, p. 108.

[59] S. HILDEBRAND, *Feinmechanische Bauelemente*, 1968, Carl Hanser Verlag, Munich.

1.3 Detachable fastenings[1–12, 16]

1.3.1 Key fastenings

Keys are rarely used in precision engineering. The hole to accommodate the key is subject to certain requirements: to ensure a good fit, the holes are reamed one above the other. In most cases a threaded structure can be used to advantage. One or two structural examples are given in the following diagrams.

The rotary (or rolling) key fastening (Fig. 1.137) is very rarely used. The other examples given refer to sliding key fastenings. For example, Fig. 1.138 illustrates the fastening of a hub to a shaft, whilst Fig. 1.139 shows a shaft fastened to a wall by means of a wedge-shaped key, that is, a tapered dowel.

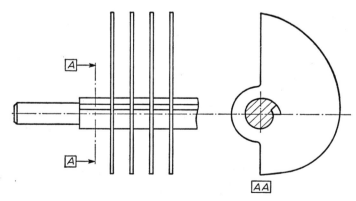

Fig. 1.137. Four capacitor plates fastened to a common spindle by means of a rolling key.

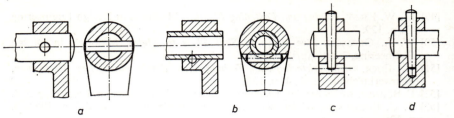

Fig. 1.138. Hub fastened to shaft by means of tapered dowel inserted as sliding key.

a. Dowel crosswise through hub and shaft.
b. Dowel passed tangentially through hub and hollow shaft.
c. Dowel emerging into drilled hole: this hole makes it easier to ream the tapered hole and withdraw the dowel.
d. Dowel protrudes, so it can be withdrawn.

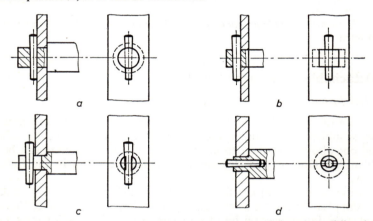

Fig. 1.139. Shaft fastened to wall by means of tapered dowel as sliding key.

a. Dowel draws round shaft in the axial direction.
b. Dowel draws flat bar in the axial direction.
c and *d.* Dowel in axial slot: end of shaft is clamped in radial direction. Shaft is anchored by friction: a matching hole is provided before milling the slot.

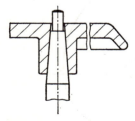

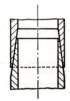

Fig. 1.140. Tapered shaft fixed in hub of disc.

Fig. 1.141. Tapered stopper, e.g. in a bottle, provides a tight seal and is firmly anchored by friction.

Fig. 1.142. Tapered pipes mated together.

Other uses are for such fastenings as: a disc on a shaft (Fig. 1.140), a stopper on a bottle (Fig. 1.141), or to join two pipes, as in the case of a vacuum cleaner (Fig. 1.142). The two parts can then be separated by twisting them relative to one another.

The fluted pins shown earlier in Figs. 1.110a and 1.110b can also be regarded as keys. The former occurs as a dowel in cylindrical holes and the latter in holes that have not been reamed out. Another example of the use of a tapered fluted pin is given in Fig. 1.143. There is limited scope for reassembling such structures after they have been demounted. Where fluted pins are used, the wall of the hole must not be too hard. The key principle as applied to clamped structures is illustrated in Figs. 1.144 and 1.145.

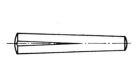
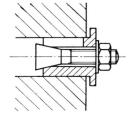
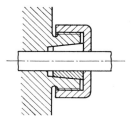

Fig. 1.143. Tapered pin with three tapered flutes: for other fluted pins see Fig. 1.110.

Fig. 1.144. Clamping structure for fastening in a hole. As the nut is tightened, the split bush is expanded by the conical member. Friction causes the bush to cling to the wall of the hole.

Fig. 1.145. Clamping structure for a shaft. As the knurled nut is tightened, the split and tapered bush is compressed into the tapered hole. The shaft is clamped in the split bush by friction.

1.3.2 Screw fastenings[13-15]

(See Volume 1, Chapter 1.8.)

There are two kinds of screw: one for fastening and the other for propulsion. The first type, the fixing screw, will be discussed here. For particulars of the different types of thread employed, see the relevant standards. Furthermore, wood screws and self-tapping screws are used where it is necessary to dispense with tapped holes. A thread, with a fully-rounded profile, is used for filament lamps (Fig. 1.146). Screws vary widely as regards type of thread, dimensions, shape of head and tip of shank. Complete data are supplied in the standards and catalogues concerned.

Screw-thread fastenings used in fine-mechanism engineering are seldom designed for strength: the torsional moment (torque) exerted in tightening the screw is usually the standard by which they are judged. Here, experience is the guide. The aim is to use as few different types of screw as possible in a given piece of equipment.

Screws made of steel (cadmium plated) or brass (nickel plated) are often employed, usually with a rolled (but sometimes with a chased) thread.

Plastics screws, mostly nylon, are preferred for electrical insulation, sealing or damping.

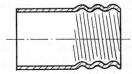

Fig. 1.146. Fully-rounded thread profile, known as an Edison thread, used in different sizes for (amongst other things) filament lamp fittings.

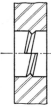

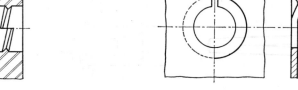

Fig. 1.147. Single turn of thread in plastics. For quick-release.

Fig. 1.148. Single turn of thread pressed in sheet material.

Internal threads can be moulded in plastics but the moulding member must afterwards be backed out. This can be avoided by moulding the bore, or core only, and tapping the thread afterwards. In other cases, only a centre for the drilling of the thread core is provided. Sometimes, a single turn of thread is sufficient (Fig. 1.147). This gives a quick-release but cannot withstand any appreciable force. A single turn of internal thread can be pressed in sheet material (Fig. 1.148). The slot is a necessary feature of the production process.

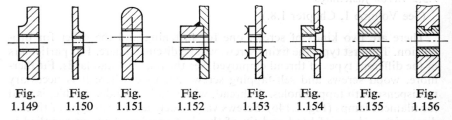

Fig. 1.149   Fig. 1.150   Fig. 1.151   Fig. 1.152   Fig. 1.153   Fig. 1.154   Fig. 1.155   Fig. 1.156

Fig. 1.149. Boss raised on wall of casting to give added length to screw-thread.
Fig. 1.150. Tapped, dimpled hole in sheet material.
Fig. 1.151. Sheet material folded over to double the length of thread.
Fig. 1.152. Nut soldered-on.
Fig. 1.153. Weld nut, welded-on. (See also Fig. 1.5b.)
Fig. 1.154. Insert nut, riveted-on: must be locked to prevent rotation.
Fig. 1.155. Nut riveted into wall of soft material: tapped afterwards.
Fig. 1.156. Insert nut embedded in plastics product.

The threaded length of the nut must be sufficient to ensure a joint of adequate strength. The problem of increasing the number of turns of thread made possible by the wall thickness of the component has been tackled in various ways. A boss can be formed on castings (Fig. 1.149). Tapped holes in sheet material are usually dimpled (Fig. 1.150). Sometimes the sheet is folded over to double the depth of the "nut" (Fig. 1.151). Nuts can also be soldered-on (Fig. 1.152), welded-on (Fig. 1.153) or riveted-on (Fig. 1.154). Special inserts are used to reinforce relatively soft materials (Figs. 1.155 and 1.156).

In all cases the nut must be locked to prevent it from rotating with the screw.

Cheese-head screws are widely used, but raised cheese-head screws improve the appearance of the product more. A round (hemispherical) head is used where a cheese-head might cause snagging.

The countersunk head is used where a flush surface is required, whilst a raised countersunk head is more decorative. Hexagonal heads, slotted or plain, are used to withstand greater forces (see Figs. 1.157 to 1.162 inclusive).

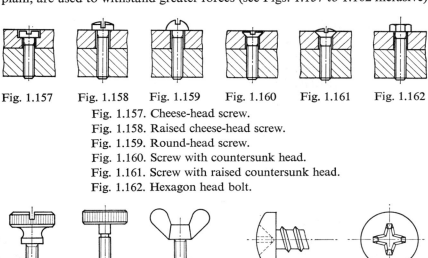

Fig. 1.157   Fig. 1.158   Fig. 1.159   Fig. 1.160   Fig. 1.161   Fig. 1.162

Fig. 1.157. Cheese-head screw.
Fig. 1.158. Raised cheese-head screw.
Fig. 1.159. Round-head screw.
Fig. 1.160. Screw with countersunk head.
Fig. 1.161. Screw with raised countersunk head.
Fig. 1.162. Hexagon head bolt.

   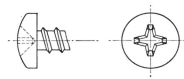

Fig. 1.163   Fig. 1.164   Fig. 1.165   Fig. 1.166

Fig. 1.163. Knurled screw with collar and slotted head.
Fig. 1.164. Knurled screw without collar.
Fig. 1.165. Wing bolt.
Fig. 1.166. Screw with cross-recessed head.

Knurled heads, with or without a collar (knurled screw) and wing heads for tightening and relaxing by finger-grip, are also available. A cross-recessed head centres the screwdriver accurately and is used on self-tapping screws. Some of these different types of head are shown in Figs. 1.163 to 1.166 inclusive.

Grub screws are headless, and simply provided with a slot or a hexagon recess. Because the thread extends to slot level, the slot tends to break out of them. On set screws, the thread ends a short way below the slot, so they are better in this respect, but cannot be sunk completely in the tapped hole by reason of the smooth head thus formed. An ordinary cheese head is very much stronger (Figs. 1.167 to 1.170 inclusive).

Screw shank ends in current use are: rounded, to bear on, say, a flat surface; conical with rounded tip, for fastening on a shaft with a V-groove, and cupped, for fastening on, say, a round shaft.

Studs are headless and may be fully-threaded (Fig. 1.171*a*) or include a cylindrical, threadless shank (Fig. 1.171*b*). These two forms of stud can be left in the tapped hole, whilst the joint members are demounted by removing a nut.

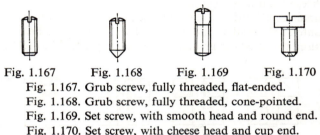

Fig. 1.167   Fig. 1.168   Fig. 1.169   Fig. 1.170

Fig. 1.167. Grub screw, fully threaded, flat-ended.
Fig. 1.168. Grub screw, fully threaded, cone-pointed.
Fig. 1.169. Set screw, with smooth head and round end.
Fig. 1.170. Set screw, with cheese head and cup end.

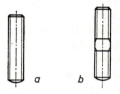

Fig. 1.171

a. Stud, fully-threaded.   *b.* Stud, partly threaded.

Where *separate nuts* are employed, they are almost invariably of steel (cadmium plated) or brass (nickel plated), hexagonal, and tightened or held by means of a spanner. Nuts held by the structural member itself should preferably be square.

Pressed nuts are cheap, but tend to lack smoothness or freedom from burr, and are therefore rarely employed. A square or hexagon nut made of thin sheet with upright edges is sometimes used for inexpensive lightweight structures. Cap nuts are neat and attractive, but expensive; however, they do obviate snagging behind a projecting end of thread. Round, slotted nuts are used where very little space is available. Round nuts, with two or more axial or radial holes in them, are slightly larger in diameter than the slotted nuts. The holes are used to tighten the nuts or hold them. Knurled nuts, with or without a collar, can be tightened with the fingers. Where a knurled nut does not provide sufficient grip, a wing nut is used instead (see Figs. 1.172 to 1.181 inclusive).

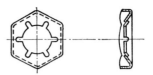

Fig. 1.172   Fig. 1.173   Fig. 1.174

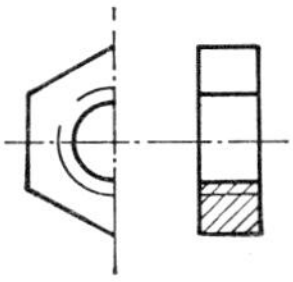

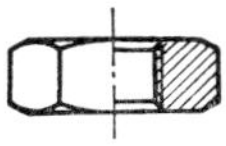

Fig. 1.172. Turned hexagon nut.

Fig. 1.173. Pressed hexagon nut.

Fig. 1.174. Nut made of thin sheet material.

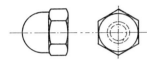

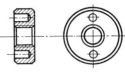

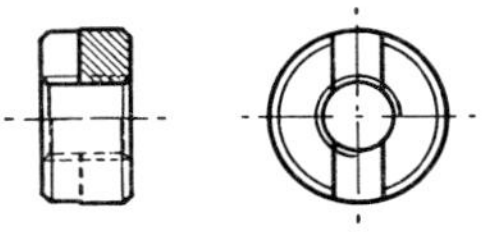

Fig. 1.175   Fig. 1.176   Fig. 1.177

Fig. 1.175. Cap nut.

Fig. 1.176. Slotted nut.

Fig. 1.177. Nut with axial blind holes for tightening.

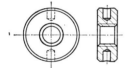

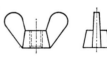

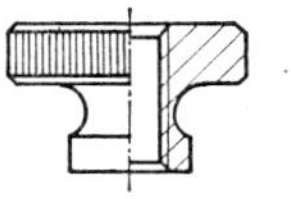

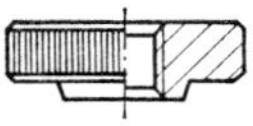

Fig. 1.178   Fig. 1.179   Fig. 1.180   Fig. 1.181

Fig. 1.178. Nut with radial blind holes for tightening.

Fig. 1.179. Knurled nut, with collar.

Fig. 1.180. Knurled nut, without collar.

Fig. 1.181. Wing nut.

Where a screw-thread is tapped in a soft material, such as zinc or aluminium alloys, to accommodate screws which have to be tightened and relaxed a number of times, the tapped hole may wear. The hole can then be enlarged, tapped again with a special tool and reinforced with a helical insert, or heli-coil, obtainable through commercial channels (Fig. 1.182). The heli-coil is made of a hard material and adapts to the screw, which can be screwed in and out without difficulty.

Screw-heads must seat cleanly. This does not occur with countersunk heads, and in addition, they give rise to two other defects: the deviation indicated by dimension "a" in Fig. 1.183, and the possibility that the recess for the screw-head may be out of line with the shank, or in other words, not properly centred. Where there is more than one countersunk screw, it is possible that both these defects will occur. Neither is encountered when the heads are seated flat, so it is better to use cheese-heads, if possible, recessed into blind holes, when a flush surface is required (Fig. 1.184). Blind holes are expensive items, however.

Bevelled washers should be used to mate screw heads to sloping surfaces (Fig. 1.185).

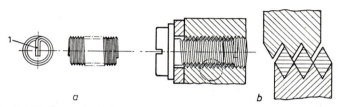

Fig. 1.182. Screw in helical insert (or heli-coil) to reinforce the wall of the tapped hole.

a. Heli-coil; 1 = tab for screwing heli-coil into hole.
b. Tightened screw in inserted heli-coil.

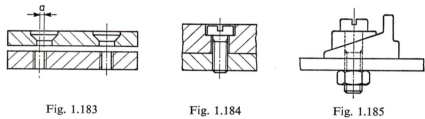

Fig. 1.183   Fig. 1.184   Fig. 1.185

Fig. 1.183. Situation in the case of a fastening involving two countersunk screws. The difference in pitch "a" is not conducive to a sound joint.

Fig. 1.184. Fastening with cheese-head screw. The head is sunk into a blind hole, so differences in pitch can be accommodated.

Fig. 1.185. Bevelled washer under screw-head, for fastening a sloping member.

Washers are used to obtain a more even distribution of the head pressure on a relatively soft or brittle material, and to avoid damage when the screw-head or nut is tightened on it. Brittle material is protected by washers of synthetic resin-bonded paper, pressboard or copper. Washers are also used to bridge elongated holes. Different sizes of washer are available for a given screw size.

Often, one screw is sufficient to provide a joint of the required strength. To prevent possible relative displacement or rotation of the joint members, there are several effective structures involving dowels, spring dowels, insert bushes, centring bosses and lugs, mating flanges and so on, some details of which are shown in Figs. 1.186 to 1.196 inclusive.

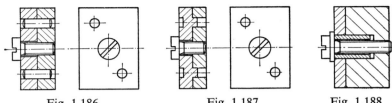

Fig. 1.186   Fig. 1.187   Fig. 1.188

Fig. 1.186. Single-screw fastening positioned by two dowels.

Fig. 1.187. Single-screw fastening positioned by embossed centring bosses in centring holes.

Fig. 1.188. Positioning by insert bush.

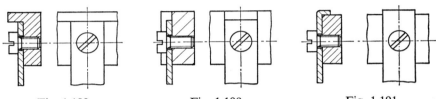

Fig. 1.189   Fig. 1.190   Fig. 1.191

Fig. 1.189. Rotation prevented by a flange.

Fig. 1.190. Rotation prevented by a (milled) slot.

Fig. 1.191. Flanged edge to prevent rotation.

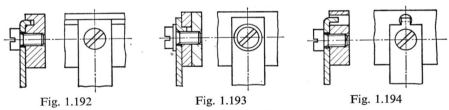

Fig. 1.192   Fig. 1.193   Fig. 1.194

Fig. 1.192. Rotation prevented by flanged edge in slot.

Fig. 1.193. Tab in cotter hole prevents rotation.

Fig. 1.194. Flanged edge in extra hole prevents rotation.

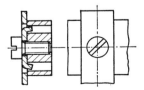

Fig. 1.195. Better location by two tabs in extra holes.

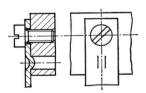

Fig. 1.196. Embossed projection in extra hole prevents rotation.

In most of the examples given, the correct location of the joint members relative to one another is also ensured, so as to cushion any shear forces that may be exerted on them.

Self-tapping screws tap the necessary (mating) thread themselves in sheet material (steel, aluminium), die castings, cast iron, thermoplastics and thermosetting plastics. The fact that they do not require pre-tapped holes is a positive advantage with cast or moulded parts. Nor do they present any problems in matching the threads of screw and nut, a further advantage where parts made in different countries need to be interchangeable.

Self-tapping screws are harder, and a little more expensive, than screws with an ordinary thread in current use.

To ensure a sound joint, it is essential to keep exactly to the prescribed hole size. Fastenings with self-tapping screws are detachable, do not shake loose easily and are much used in set manufacture where this is compatible with the accuracy required. It is advisable to employ the thread forming type of self-tapping screw (Figs. 1.197 and 1.198) in thermoplastic sheet material of moderate thickness. A cutting screw is required in thicker walls, die-castings, cast iron and thermosetting plastics, because thread forming is more difficult in them (Figs. 1.199 and 1.200).

Although all the different types of head are available, the cross-recessed head is most widely used.

| Fig. 1.197 | Fig. 1.198 | Fig. 1.199 | Fig. 1.200 |

Fig. 1.197. Self-tapping screw shank of the thread forming type. Suitable for structures of sheet material. The point facilitates initial penetration.

Fig. 1.198. Self-tapping screw shank of the thread forming type. Suitable for thick sheet, non-ferrous castings and plastics. The material deforms readily and the pre-drilled hole is larger than the root diameter of the screw.

Fig. 1.199. Self-tapping screw shank of the cutting type with a chip slot, suitable for metals of moderate strength, for plastics and for strong, but brittle, materials. Less force is required to screw them in than for the corresponding displacement type. The screw-thread resembles a machine-tapped thread and tightens well.

Fig. 1.200. Self-tapping screw shank of the cutting type with five chip slots: is less likely to shear the thread in brittle plastics and castings. Suitable for long threads in blind holes.

1.3.3 Locking screw fastenings

Methods of locking are designed to increase the torque needed to relax a screw/nut fastening. In principle, there are three possible kinds of locking, based on:

● The shape, e.g. through the use of retaining parts (positive locking).
● The action of a force, usually friction.
● A permanent joint, e.g. adhesive or welded.

The following considerations govern the choice of a suitable method of locking:

- What is the risk associated with possible failure of the fastening?
- Can the nut be tightened hard against the workpiece?
- How can the gripping or binding force be set up, and is it lasting?
- How high is the temperature of the joint and what are the climatic conditions?
- Will the fastening have to be released often?
- What is the required lifetime?
- Is the locking method fast and cheap enough?
- Is any damage caused by locking permissible?

Only a few of the possible structures will be illustrated here, such as: positive locking (Figs. 1.201 to 1.207 inclusive); locking by friction (Figs. 1.208 to 1.218 inclusive); permanent joints (Figs. 1.219 to 1.221 inclusive). The various features are clarified by the captions.

If a screw fastening has to be undone frequently, it may be advisable to assemble the screw or nut concerned in such a way as to hold it captive, so that it cannot be mislaid. Such captivation can be carried out in various ways, as demonstrated in Figs. 1.222 to 1.230 inclusive. Structures can also be rendered tamper-proof, or inaccessible to unauthorized persons, say, by methods shown in Fig. 1.231 to 1.233 inclusive. Dismantling can also be discouraged by employing a screw which requires a special tool, not readily obtainable, to unfasten it.

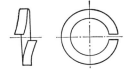 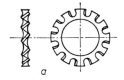

Fig. 1.201. Spring lock-washer, which locks nut (screw) when interposed between nut (screw head) and contact surface.

Fig. 1.202. Toothed spring lockwashers.

a. With outside teeth.
b. With inside teeth.

a locks better than b, but b causes no visible damage and looks better.

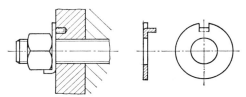 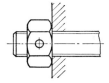

Fig. 1.203. Locking plate with raised tabs, locks nut.

Fig. 1.204. Pin, in hole drilled after assembly, locks nut.

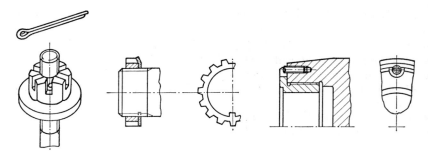

Fig. 1.205   Fig. 1.206   Fig. 1.207

Fig. 1.205. Castle-nut locked by split pin passed through hole in bolt.

Fig. 1.206. Toothed lockwasher with a tab bent into one of the slots of the nut. Because of the difference in the number of tabs and slots, one of the tabs invariably coincides with a slot.

Fig. 1.207. Threaded part, locked by grub screw.

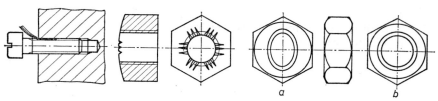

Fig. 1.208   Fig. 1.209   Fig. 1.210

Fig. 1.208. Locking strip of plastic between screw and oversize tapped thread in plastic.

Fig. 1.209. Nut with deformed threads locks by friction on thread of bolt.

Fig. 1.210. Nut, with oval bore at the top, locks by friction on thread of bolt. *a.* Top.   *b.* Bottom.

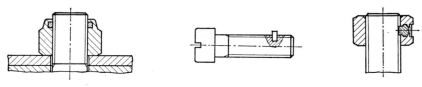

Fig. 1.211   Fig. 1.212   Fig. 1.213

Fig. 1.211. Nut with frictional (fibre) lockwasher.

Fig. 1.212. Screw with nylon insert for added friction.

Fig. 1.213. Nut with nylon insert for added friction.

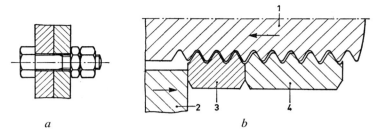

          a   *b*

Fig. 1.214. *a*. Bolt with nut and jamb nut. The full nut should be tightened on top
          of the jamb nut.

          b. The top nut (4) takes up the force (arrowed) in the bolt (1). The
          bottom one (3) acts as a washer during the initial tightening of the top
          nut on to the surface of part (2) and as a jamb nut for final tightening.

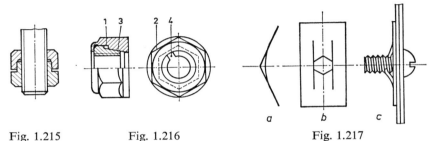

Fig. 1.215   Fig. 1.216   Fig. 1.217

Fig. 1.215. Set of jamb nuts of equal height externally, which facilitates tightening.

Fig. 1.216. Composite lock-nut in which the split internal nut is driven by the
          hexagonal recess in the outer nut. When the outer nut is tightened,
          the inner nut clamps on to the bolt thread.
          1 = hexagon;   2 = hexagon;   3 = taper;   4 = gap.

Fig. 1.217. Punched plate which acts as a self-locking nut (see also Fig. 1.242).
          a. Side view.   *b*. Top view.   *c*. Assembled.

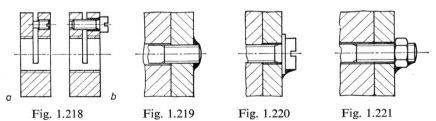

          Fig. 1.218   Fig. 1.219   Fig. 1.220   Fig. 1.221

Fig. 1.218. Split lock-nuts; when screw is tightened, the nut clamps on to the
          bolt thread.
          a. With grub screw.   *b*. With cheese-head screw.

Fig. 1.219. Stud, locked with paint.

Fig. 1.220. Screw head and washer, locked with paint.

Fig. 1.221. Bolt and nut locked with paint.

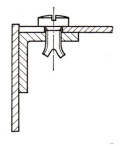

Fig. 1.222. Captive screw.

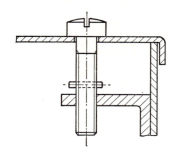

Fig. 1.223. Pin through screw makes
it captive.

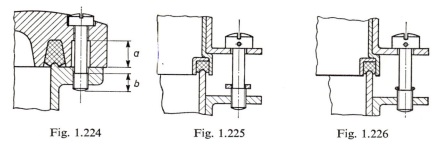

Fig. 1.224   Fig. 1.225   Fig. 1.226

Fig. 1.224. Captive screw: shank diameter is smaller than root diameter of thread:
for demounting, *a* must be larger than *b*.

Fig. 1.225. A ring of synthetic resin-bonded cardboard, pressboard or fibre,
twisted on to the thread of the screw, holds it captive.

Fig. 1.226. As Fig. 1.225, but the ring is resilient and formed of steel wire.

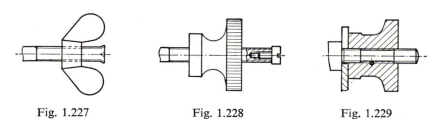

Fig. 1.227   Fig. 1.228   Fig. 1.229

Fig. 1.227. Staking the stud makes the wing-nut captive.

Fig. 1.228. Extra screw holds the knurled-nut captive.

Fig. 1.229. Pin through knurled-nut and passing narrowed stud hold nut
captive.

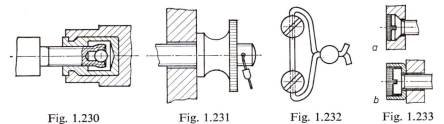

Fig. 1.230   Fig. 1.231   Fig. 1.232   Fig. 1.233

Fig. 1.230. Once the nut is tightened, the ball expands the end of the stud, thereby
holding the nut captive.

Fig. 1.231. Nut sealed with lead tag.

Fig. 1.232. Screws sealed with lead tag.

Fig. 1.233. Sealed fastening.
        a. Countersunk screw sealed with paint.
        b. Cheese-head screw sealed with paint.

1.3.4 Special screw fastenings

A bayonet fastening is obtained by inserting one member in the other
and twisting them (through a small angle) relative to each other. The
fastening can be released quickly by reversing the process. Bayonet fastenings
can be locked by friction between the members, by spring action or by keying.
Structural examples are given in Figs. 1.234 to 1.239 inclusive.

Other fastenings which can be released by relaxing the fixing element
slightly are marketed in the U.S.A. by companies specializing in the manufacture of fasteners. One or two products of this kind are shown in Fig. 1.240
and Fig. 1.241.

Punched nuts with a single turn of thread can be augmented with special
parts, enabling them to be assembled from one side or holding them captive.
They also enable other structural members to be fastened to (or located on)
them. There is a wide range of possibilities (Fig. 1.242).

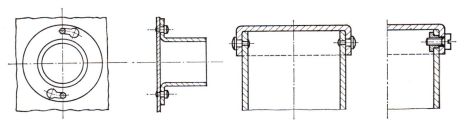

Fig. 1.234. Bayonet fastening of flange
to plate.

Fig. 1.235. Cover locked on can by
bayonet fastening.

3*

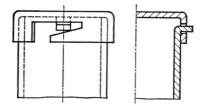

Fig. 1.236. As Fig. 1.235, but with formed tabs.

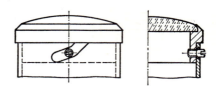

Fig. 1.237. Lens assembled in fitting.

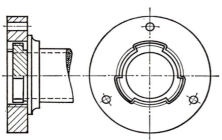

Fig. 1.238. Pipe joined to flange: screw thread has been removed in three places for quick-release.

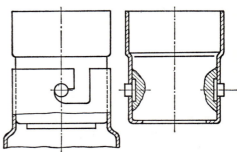

Fig. 1.239. Interlocked bayonet fastening.

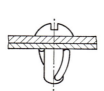

Fig. 1.240. "Spring-Lock" made by Simmons Fastener Corp. After the fastener has been inserted and turned a quarter-turn, the U-shaped spring steel wire, passed through the shank of the pin, provides the locking force. Another quarter turn in the same direction releases the fastener.

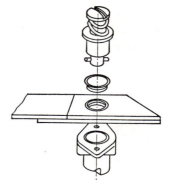

Fig. 1.241. "Camloc" made by Camloc Fastener Corp. As the fastener is turned a quarter-turn, the pin passed though the stud shank tightens the fastening.

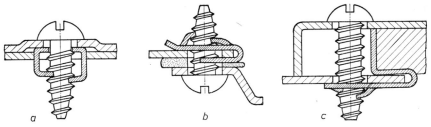

Fig. 1.242. Punched nuts (or "speed nuts" as they are called) with a single turn of thread, augmented by additional parts.

a. Suitable for blind assembly. When the screw is tightened, the U-shaped nut grips into the bottom plate.

b. The clip on the nut is slipped on to the top plate and held in place by hooking back.

c. The clip on the nut is slipped on to the bottom plate: the upright edge forms a stop for the plate between.

REFERENCES

[1] RICHTER, V. VOSS and KOZER, *Bauelemente der Feinmechanik*, Verlag Technik, Berlin, 1954.
[2] K. HAIN, *Die Feinwerktechnik*, Fachbuch Verlag Dr Pfanneberg, Giessen, 1953.
[3] G. SCHLEE, *Feinmechanische Bauteile*, Verlag Konrad Wittwer, Stuttgart, 1950.
[4] D. C. GREENWOOD, *Product Engineering Design Manual*, McGraw-Hill, New York/Toronto/London, 1959.
[5] K. RABE, *Grundlagen Feinmechanischer Konstruktionen*, Anton Ziemsen Verlag, Wittenberg/Lutherstadt, 1942.
[6] R. PÖSCHL, *Verbindungselemente der Feinwerktechnik*, Springer Verlag, Berlin/Göttingen/Heidelberg, 1954.
[7] H. EDER and W. P. UHDEN, *Taschenbuch der Feinwerktechnik*, C. F. Winter'sche Verlaghandlung, Prien, 1965.
[8] *Fasteners for Packaging/Production*, Special Fastener Supplement of Electronic Packaging and Production, volume 6, section 2, 1966, September.
[9] *Fastening and Joining Reference Issue*, Machine Design, 1967, June 15.
[10] F. WOLF, *Die Schrauben und Keilverbindungen in der Feinwerktechnik*, Deutscher Fachzeitschriften- und Fachbuchverlag, Stuttgart, 1955.
[11] *The Fasteners Book*, Machine Design, 1960, September 29.
[12] J. SOLED, *Fasteners Handbook*, Reinhold, New York, 1957.
[13] S. H. DAVISON, *How to select Threaded Inserts*, 1961, May 22, p. 46.
[14] Catalogue JEVEKA No. 35, Amsterdam.
[15] Miniatuurschroefdraad ISO, Toleranties NEN 1921; Nominale maten NEN 1920.
[16] S. HILDEBRAND, *Feinmechanische Bauelemente*, 1968, Carl Hanser Verlag, Munich.

1.4 Guide systems (slideways, linear bearings, etc.)[1 – 6, 8, 14]

1.4.1 Introduction

Some structural elements, unlike fastenings, permit a specific type of motion. The possibilities of relative motion between two bodies can be expressed in terms of degrees of freedom. In all, there are six possible degrees of freedom, namely: translation in the three co-ordinate directions (at right-angles) x, y and z, and rotation about the three co-ordinate axes u, v and w. In the case of a bearing, translation, in any direction, is prohibited, leaving freedom to rotate on one axis only. Structural elements called guides (slideways, linear bearings) have, at most, two degrees of freedom: one motion of translation and one of rotation. A structural example of this is given in Fig. 1.243a; if translation of this cylindrical guide is prevented, only one degree of freedom remains: the possibility of translation (Fig. 1.243b and c).

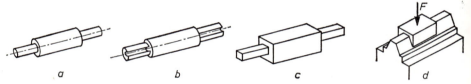

Fig. 1.243. Survey of different types of guide.

a. Closed cylindrical guide, rotation unobstructed (two degrees of freedom).
b. Closed cylindrical guide, with obstruction to prevent rotation (one degree of freedom).
c. Closed angular-section guide (one degree of freedom).
d. Open angular-section guide. Force F is required to hold the two members together.

A guide, then, is a structural element enabling one member to slide along another in a given path. In a linear guide, the path is rectilinear. In its simplest form the path (or way) is cylindrical or angular in cross-section. The movable part, or slide, may fully or partially enclose this slideway (closed or open guide).

In the case of an open guide (Fig. 1.243d) a certain force F is required to hold the two members together. Angular-section guides are usually open, and cylindrical guides lend themselves to construction in the closed form, as seen in Figs. 1.243a and b. Some cylindrical guides include a device to prevent rotation, where necessary.

A distinction is usually made between sliding and rolling guides. What is, in effect, a straight guide can also be obtained by a system of levers. The structural elements required for this purpose are bearings and connecting rods (see also Chapter 2 and Volume 1, Chapter 3).

For particulars of lubrication, see Volume 8, Chapter 5.

Lubricants are described in Volume 2, Chapter 9.

1.4.2 Sliding guides[7, 12]

An ordinary guide may stick (or jam) (Volume 1, Section 2.3.8). In the case illustrated in Fig. 1.244a, friction is ruled out, in theory, if the force F acts directly on the centre-line of the rod, since no pressure is exerted at right-angles between rod and guide. Jamming is therefore out of the question. This can be demonstrated as follows

$$N_1 = N_2 = \frac{F.r}{l}$$

follows from the condition of equilibrium with respect to rotation.

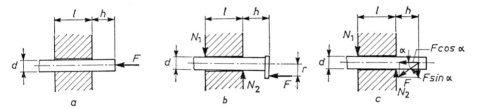

Fig. 1.244. Case of a straight guide subject to jamming.

a. The sliding force F acts directly on the rod.
b. The sliding force F acts off-centre.
c. The sliding force F acts at an angle to the direction of sliding.

In sliding, the rod is subjected to the frictional force

$$\mu(N_1 + N_2) = \frac{\mu.2F.r}{l}$$

where μ = the coefficient of friction.

Jamming does not occur if the frictional force is less than the sliding force or thrust F. Therefore

$$\frac{\mu.2F.r}{l} < F \quad \text{or when} \quad \frac{l}{r} > 2\mu.$$

Frictional forces also occur in the case illustrated in Fig. 1.244c, and may cause jamming.

To demonstrate this

$$N_1 = \frac{F \sin \alpha.h}{l} \quad \text{and} \quad N_2 = F \sin \alpha . \frac{h + l}{l}$$

Thus, if d is very much smaller than h, the component $F \cos \alpha$ must overcome the two frictional forces before sliding begins. Therefore

$$F \sin \alpha > F \sin \alpha . \frac{2h + l}{l} \mu \quad \text{or} \quad \frac{1}{\tan \alpha} > \frac{2h + l}{l} \mu$$

If

$$\frac{l}{h} > \frac{2\mu \tan \alpha}{1 - \mu \tan \alpha}$$

sticking does not occur.

A guide in proper working order must be incapable of sticking. To achieve this, due consideration should be given to the following points:

- The thrust should be concentrated as much as possible along the path. In other words, it should exert the least possible moment on the guiding member.
- The members will slide more easily relative to one another if the coefficient of friction between them is low.
- A long enclosure helps the slide to run lightly.

Important features are: the focus of action and the direction of the thrust, the length of enclosure and the value of the coefficient of friction. The clearance and the diameter (or the equivalent dimension of angular guides) are less important from the point of view of possible frictional interference or sticking.

For accurate positioning, however, the clearance must be kept small, as also to avoid undue surface pressure (a high coefficient of friction accompanied by seizing) at the end of the enclosure.

At the same time the clearance should not be too small to allow:

- Possible deformation in motion.
- Deviations of form resulting from the manufacturing process.
- Room for the lubricant, for corrosion products and for expansion due to heating or the absorption of moisture.

Possible spontaneous sliding can be counteracted by applying a (adjustable) frictional force.

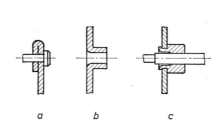

Fig. 1.245. Methods of lengthening the guide enclosure.

  a. Double wall-thickness.
  b. Extruded cylindrical hole.
  c. Bush riveted on.

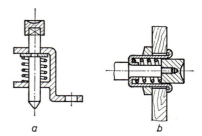

Fig. 1.246. Practical version of a double guide in a wall, with a compression spring to return the sliding member.

  a. Two flanges with holes for the guide.
  b. Specially-fitted bush, with one enclosure for the shaft, another for the arbor.

Various designs are used to lengthen the enclosure, which is particularly necessary when the guide has to pass through a relatively thin wall (Fig. 1.245). The length of the enclosure can be much increased by placing two guides some distance apart. This is the basis of the structures illustrated in Fig. 1.246 and Fig. 1.247.

Closed *cylindrical* guides are sometimes made of seamless pipe or rolled sheet. Such tubular guides occur in stands, optical instruments and so on. Fig. 1.247*d* shows a guide for tube supports which does not make contact over its full length, but only via indented ribs. The diameter of the rib governs the clearance. The ribs also act as stops, and form a dust-proof seal. Other methods of eliminating play and setting-up a certain amount of frictional force are illustrated in Fig. 1.248. Here, the tabs provided in the inner tube are bent outwards slightly on assembly.

Very strict accuracy is too much to expect of a guide. A strip of fabric or felt must be introduced to exclude light from the guide where necessary (Fig. 1.249). The sliding member can be prevented from twisting in the

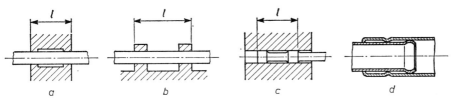

Fig. 1.247. Long enclosures obtained by placing two guides some distance apart (double guide). *l* is the length of the enclosure.

<ul>
<li>*a.* Through a wall.</li>
<li>*b.* On a casting.</li>
<li>*c.* Piston-shaped rod.</li>
</ul>

d. Guide for tubular stands made of precision-drawn brass pipe and provided with dust seal and stop.

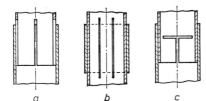

Fig. 1.248. Elimination of play and provision of necessary friction in tubular guide.

a. Four tabs bent outwards slightly on assembly.

b. Two cut-out strips for the same purpose are stronger, but less resilient.

c. The two spring tabs lack strength.

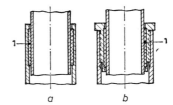

Fig. 1.249. Guide with air-tight seal.

a. Simple version.

b. Structure with bush screwed in.

1 = strip of fabric coated with adhesive.

manner shown in Fig. 1.250. As a general rule, the obstruction to prevent
rotation should be taken outwards as far as possible (long lever, therefore
small angular displacement). Guides with flexible members are required
where a force or motion has to be transmitted subject to variation of direction
or position of the ingoing (relative to the outgoing) motion of an instrument.

Figures 1.251 and 1.252 show a press-on tool (pressure applicator) for dental
work and a shutter release for a camera, respectively. The principal features
of both these constructions are a sliding core and a flexible tube, or conduit.
The core may also be a length of wire or cable stressed by a spring (Bowden
cable). The conduit is usually anchored to the frame, whilst the core slides;
a stationary core with the tube sliding on it is also possible, however, and
is sometimes preferred for structural reasons.

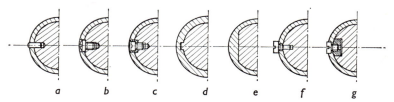

Fig. 1.250. Means of preventing the sliding member from rotating. As well as a
recess in the tube, and a pin pressed into the rod as at *a* or a screw driven in *b*
for light loads, there are: a slide block screwed on *c* for heavier loads and *d* and *e*
for heavier loads or faster speeds. The enclosing member can also be made up of
two blanks set some distance apart. *f* and *g* are the opposite of *b* and *c*.

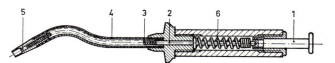

Fig. 1.251. Pressure applicator for dental work, with flexible guide. Plunger (1)
imparts the motion via rod (2) to the flexible but axially-rigid core of wire wound
into a helix (3) in the flexible conduit (4), causing the ram (5) to emerge. The
spring (6) returns the parts displaced.

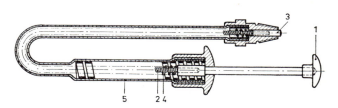

Fig. 1.252. Shutter release for camera, with flexible guide. Plunger (1) drives pin (3)
via the axially-stiff core (2). The core is guided in the helical spring (4), which
returns the parts displaced. Conduit (5) encloses the system. The structure is
stiffened at the plunger end by a helical spring made of strip.

It is also possible to combine the two designs. The flexible conduit and the play between core and conduit result in backlash. Backlash can be limited by:

- Employing conduit in the form of (phosphor bronze) wire wound into a helix under axial pre-stress. Such conduit is not flexible in the axial direction.
- Employing a core similarly constructed, if it is to be subjected to compressive loads only, or of straight (steel) wire for exclusively tensile loading.
- So arranging matters that the core only comes in contact with one side of the conduit.

The combination of tension-compression and rotation is difficult to achieve in one flexible conduit.

Angular guides are found, amongst other things, in sliding gauges (Fig. 1.253) and slide rules, where spontaneous sliding is prevented by frictional resistance set up by the spring included to eliminate play.

Fig. 1.254 illustrates a simple method of giving a strip some scope for sliding. An open guide exploiting gravity is shown in Fig, 1.255: the sliding member can be locked temporarily by means of the screw. In Fig. 1.256, the vee-shaped sliding member is guided along a cylindrical rod, whilst rotation

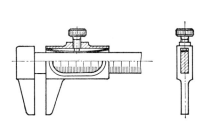

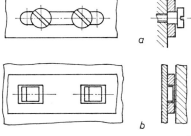

Fig. 1.253. Closed angular guide of a sliding gauge. The spring eliminates play and at the same time provides the necessary friction. Turning the knurled screw locks the gauge.

Fig. 1.254. Strip containing elongated hole is guided by:
    a. Two shouldered screws.
    b. Two tabs bent upwards.

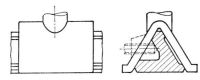

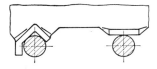

Fig. 1.255. Open angular guide, without play, for optical instruments.

Fig. 1.256. Open guide, without play, having a flange to prevent rotation along two cylindrical rods.

is prevented by a flat surface of moderate length on the slide, which travels along another rod.

The motion is free from play, even after a certain amount of wear has occurred. Some lack of parallelism on the part of the rods is permitted.

1.4.3 Rolling guides[9, 10, 11, 13]

When the guide is required to be very light-running, rolling friction is called for. The criteria for sticking, or frictional interference, are as valid here as for sliding guides, the only difference being that rolling guides are less likely to stick, by reason of their very much lower coefficient of friction.

The rollers may be spherical, cylindrical or in the form of a diabolo. Unlike sliding guides, they permit a certain amount of pre-stress in assembly without adding unduly to the thrust required to set them in motion.

Cylindrical rolling guides are shown in Fig. 1.257. In its simplest form (Fig. 1.257a with three rollers) the structure permits unlimited linear travel, unlike the hardened bushing with balls (Fig. 1.257b), which only allows the shaft a limited amount of sliding motion. Although the balls revolve freely in the slots in the bushing, they cannot drop out of them. In the composite guide bush (Fig. 1.257c) with three ball-races, or tracks, in which the balls can circulate, a shaft has unlimited linear travel.

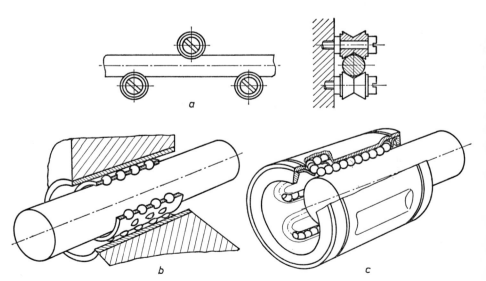

Fig. 1.257. Closed cylindrical rolling guide for shafts, consisting of:
    a. three diabolo-shaped rollers.
    b. a hardened bushing with roller balls.
    c. a complete guide bush containing three ball tracks.

The next four illustrations (Figs. 1.258–1.261 inclusive) show examples of guides along rods of angular section.

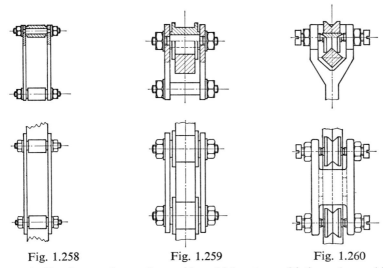

Fig. 1.258   Fig. 1.259   Fig. 1.260

Fig. 1.258. Simple, angular-section guide, which sets up friction when subjected to transverse loads.

Fig. 1.259. Similar to Fig. 1.258, but the transverse forces are absorbed by flanges on the rollers.

Fig. 1.260. Angular-section rolling guide using vee rollers: no play.

Figure 1.258 is a simple design, but it sets up a great deal of friction when acted on by transverse forces. In Fig. 1.259, friction occurs between the flanges. The guide in Fig. 1.260 is without play, but, owing to the difference in peripheral velocity between different points on the surface of the roller, the motion is not strictly confined to rolling.

The system of rollers staggered in pairs, shown in Fig. 1.261a, provides purely rolling motion without play. A staggered roller guide, as used in typewriters, is similarly constructed (Fig. 1.261b).

The rollers ride in line and are held in place by a cage. Roller strips of many different types and sizes made of plastics, amongst other things, and with retained rollers so that they do not drop out, are available through commercial channels (Fig. 1.261c). A roller chain (Fig. 1.261d) can follow a curvilinear path. It consists of cage elements, each containing a roller, fastened together but free to rotate. A roller chain permits unlimited motion. The problem of accurate tracking without twisting has been solved, for a carriage, in Fig. 1.262. Where there are forces acting in different directions, the closed guide of Fig. 1.263 can be used instead. Similar open (or closed) guides can be constructed of roller strips fitted with rollers, balls or needles (Fig. 1.264).

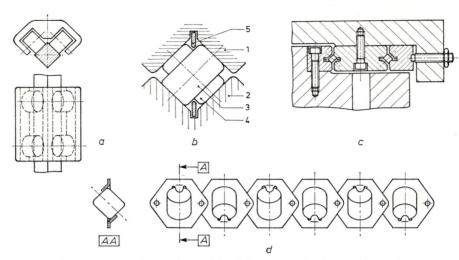

Fig. 1.261. Angular-section guide with staggered rollers, without play.

a. Rollers riding abreast on square rod.
b. Rollers riding in-line between carriage and frame.
    1 = carriage. 2 = frame. 3 = front roller. 4 = rear roller. 5 = cage.
c. Slide with assembled roller strips.
d. Roller chain.

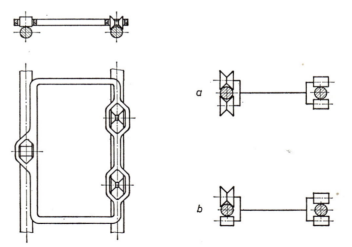

Fig. 1.262. Accurate tracking of carriage by means of two vee-rollers and
one cylindrical roller on two round
rods.

Fig. 1.263. Structure similar to that of
Fig. 1.262, for use where forces are
exerted in more than one direction.
b is a slightly simpler version of a.

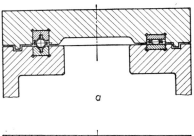

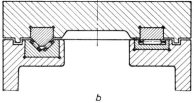

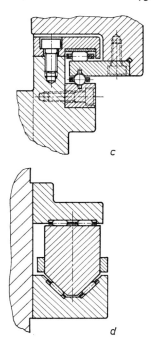

Fig. 1.264. Guides constructed of roller strips.

a. Open guide on balls and rollers.
b. Open guide on needles.
c. Closed guide on balls and rollers.
d. Closed guide on needles.

REFERENCES

[1] RICHTER, V. VOSS and KOZER, *Bauelemente der Feinmechanik*, Verlag Technik, Berlin, 1954.
[2] K. HAIN. *Die Feinwerktechnik*, Fachbuch Verlag Dr. Pfanneberg, Giessen, 1953.
[3] G. SCHLEE, *Feinmechanische Bauteile*, Verlag Konrad Wittwer, Stuttgart, 1950.
[4] D. C. GREENWOOD, *Product Engineering Design Manual*, McGraw-Hill, New York/Toronto/London, 1959.
[5] K. RABE, *Grundlagen Feinmechanischer Konstruktionen*, Anton Ziemsen Verlag, Wittenberg/Lutherstadt, 1942.
[6] F. WOLF, *Lagerungen, Geratführungen und Kupplungen in der Feinwerktechnik*, Deutscher Fachzeitschriften- und Fachbuchverlag, Stuttgart, 1956.
[7] K. HOLECEK, *Über das Verkannten*, Feinwerktechnik, 1956, 10, p. 353.
[8] G. REIJNGOUD, *Geleidingen voor heen-en-weergaande bewegingen*, De Constructeur, 1962, January/February.
[9] *Kraftverteilung in Wälzkörpergelagerten Geradführungen*, Konstruktion, 1960, 9, p. 353.
[10] K. MÜLLER, *Wälzkörpergelagerte Längsführungen*, Werkstatt und Betrieb, 1958, 2, p. 73.
[11] *Längsführungen*, Katalogus W. Schneeberger, Roggwil, Switzerland.
[12] J. M. EASTMAN, *Designing for Less Friction*, Product Engineering, 1952, May, p. 124.
[13] J. A. HOPE, *Rolling-Element Linear-Motion Bearings*, Bearings Reference Issue, Machine Design, 1966, March 10.
[14] S. HILDEBRAND, *Feinmechanische Bauelemente*, 1968, Carl Hanser Verlag, Munich.

1.5 Bearings[1-9, 42]

1.5.1 Introduction

Bearings, like guides, usually have only one degree of freedom. These structural elements allow only rotary motion, which may involve the transmission or absorption of radial and/or axial forces. Hence, there are radial bearings and axial (or thrust) bearings: also many bearings capable of withstanding both radial and axial forces. A variety of combined radial/axial load bearings are used in precision engineering or fine mechanisms, as they may be called.

In this section they are classified according to the way friction is set up in the bearing. Thus, the following types of bearing have been established:

(a) *Plain bearings*, having the shaft (journal) and the bearing (metal) in direct contact with one another, with or without lubrication. They can be subdivided according to the shape of the sliding contact-surfaces, as follows:

● Journal bearings.
● Cone bearings.
● Cone bearings with spherical journal.
● Pivot bearings (cupped bearing surface with pointed journal).

(b) *Anti-friction bearings*, in which the members rotating relative to one another are separated by revolving elements, which may be round (balls), cylindrical, tapered, barrel-shaped or needle-shaped.

Axial types of ball bearing, cylindrical roller bearing and needle roller bearing exist. Taper roller bearings and spherical (or barrel-shaped) roller bearings can absorb radial forces as well as axial forces. The same applies to one or two types of ball bearing.

Ball bearings are the type of anti-friction bearing most used in fine mechanisms: cylindrical roller bearings are employed occasionally, but taper roller bearings, spherical roller bearings and needle roller bearings, whose primary function is to withstand heavy loads, are hardly ever used for fine mechanisms.

There are also miniature anti-friction bearings (in which the outside diameter of the external race is less than 10 mm).

Instrument anti-friction bearings are miniature bearings satisfying one (or both) of the following requirements:

● Very true running.
● A minimum of friction.

Combinations of radial and axial anti-friction bearings are employed in a very wide variety of structures.

(c) *Special bearings*

Knife-edge bearings, with a vee-shaped knife-edge, supported by a seat in such a way that it can pivot to-and-fro through a small angle.

Magnetic bearings, based on the mutual repulsion or attraction of built-in magnets.

Tension strip bearings, capable of limited rotary motion through the elastic torsion of a tension strip.

Air bearings, in which gas friction occurs.

Hinge springs, for particulars of which see Section 1.7.5.

Some bearings remain stationary, whilst the shaft rotates: in others, the shaft remains stationary.

The shape of the shaft (and journal) is often governed by the process used for manufacturing the members to be joined by the shaft. Shafts may be supported at one end or at both ends. The angle of rotation is often limited, as in structures for adjustment, control, switching and pivoting. The need to withstand (minor) axial forces is invariably taken into account in the design. Usually, the size of a bearing in precision engineering is established on the basis of reliability in service (also after transport) and the method of manufacture, rather than on permissible surface pressure, strength or possible dissipation of heat.

Precise calculation is necessary, however, to cope with high speeds or a very small shaft diameter. Particulars of these calculations are to be found in conventional engineering textbooks.

Bearing materials are discussed in Volume 2, Chapter 1.6, and lubrication methods in Volume 8, Chapter 5.

1.5.2 Plain cylindrical bearings[10 – 14, 32, 33, 37, 41]

For details of the practical application of bearing materials and lubrication, see Volume 8, Chapter 5. The problem of lubrication often receives scant attention when the motion is not continuous, or the peripheral velocity of the journal is small.

It is considered sufficient to choose different materials for the bearing members, so as to avoid local friction welding (seizing) of the bearing material to the journal, The combinations of steel and brass, steel and bronze, steel and cast iron are often used, as also steel and jewels in instruments and clocks.

The bearing usually surrounds the journal completely, but where the journal is not completely enclosed, a force (a weight or a spring) must be employed to hold it in the bearing. Most bearings are cylindrical. The usual arrangement is for the bearing to stand still and the shaft to rotate. Shafts can be enclosed in various ways (Section 1.5.5). When an ordinary cylindrical bearing bore in the existing wall (plate) will not suffice, various forms of bearing-bush can be inserted: they can be anchored by forcing in, embedding, riveting, flanging or screwing (Figs. 1.265 to 1.268 inclusive).

The bearing may be recessed in a screw (Fig. 1.269). But, because of the relative inaccuracy of the screw-thread fastening, which may drag the play in the bearing to one side, this construction may easily result in misalignment. When this drawback is acceptable, a cheap bearing can be obtained thus, with scope for adjustment.

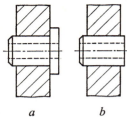

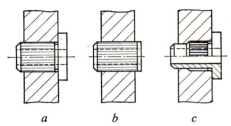

Fig. 1.265. Smooth bearing-bush forced into wall. Accurate fitting is necessary.

a. Bush with collar as axial stop.
b. Bush without collar. This is difficult to position correctly.

Fig. 1.266. Knurled bearing-bush forced into wall. The fit need not be as accurate.

a. Bush with collar as axial stop.
b. Bush without collar. This is difficult to position correctly.
c. Knurled bush with smooth lead-in centres better in the hole.

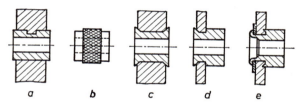

Fig. 1.267. Bearing-bushes introduced into wall in various ways.

a. Embedded bush, anchored by slot.
b. Bush for embedding, with diamond knurl to anchor it.
c. Riveted bush: hole reamed afterwards.
d. Flanged bush: hole reamed afterwards; to prevent twisting see Fig. 1.66.
e. Flanged bush that does not require reaming: preferred shape of flange. Bush is turned from hexagonal material. The points thus obtained bite into the soft wall and prevent rotation.

Fig. 1.268. Bush screwed to wall.

a. Bush with flange fastened by three screws driven into tapped holes in wall.
b. As *a*, but with screw-thread in extra ring, if wall is unsuitable for tapping.
c. Front view of extra ring used in *b*.

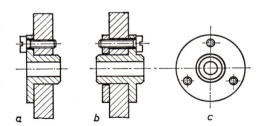

Jewels are special bearings, used in clocks, etc. They are obtainable in various shapes, and are gripped in a brass setting by indentation or clinching. For practical examples of their use see Fig. 1.270, and for further particulars see Volume 3, Chapter 4.2.

The holes in jewels may be either straight or olive-shaped.

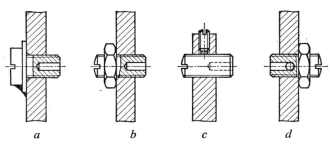

a   *b*   *c*   *d*

Fig. 1.269. Position of bearing is readjustable by means of a screw, but this introduces some inaccuracy in regard to location and alignment of the bearing centre-line. Bearings shown are:

a. In cheese-head screw: position of bearing is determined by a washer under the head, which can be locked by applying paint.
b. In grub screw: locked by a nut.
c. In grub screw: locked by a second grub screw, with a soft copper plug below it to prevent damage to the thread.
d. In grub screw: locked by a nut. A ball is pressed into the bearing recess to absorb axial forces (or thrust) with a minimum of friction.

Fig. 1.270. Jewel bearings in a brass setting.

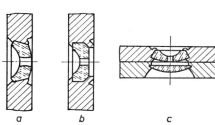

a. Bearing with olive hole and oil sink.
b. Bearing with straight hole and oil sink.
c. As *a*, fitted with end stone to absorb thrust.

a   *b*   *c*

The olive-shaped holes have a slightly dished shape that gives scope for a certain amount of angling, whilst the oil in the lubricating chamber (or oil sink) infiltrates to the position of least play.

For particulars of *plastics bearings* see Volume 10, Section 5.5.6. Polyamide and polyacetal are mainly used for this purpose. Lubrication can be dispensed with where necessary as, say, in textile processing machinery. Due consideration must be given to the lack of thermal conductivity and the appreciable thermal expansion of the material, and to the fact that it swells on absorbing moisture. Steel is chosen as the shaft material. Copper and aluminium alloys can cause excessive wear because of catalytic action.

The frictional torque exerted on a bearing journal (Fig. 1.271) is the product of the radial force (F) acting on the bearing, the coefficient of friction (μ) and the radius (r) of the journal. Formulated:

$$M = \mu F . r = W . r$$

where $W =$ the frictional force.

For free-running shafts, then, not only the force and the coefficient of friction, but also the journal radius, must be small. Accordingly, polished or rolled journals of the smallest diameter are employed. They are usually integral with the shaft, as will be seen from the examples in Fig. 1.272. The bevelled edge at a reduces the frictional moment set up by an axial force acting on the shaft, by reducing the leverage available to the frictional force. This effect is put to very good use when the end of the journal is rounded to absorb thrust (Figs. 1.272b and c). Very thin journals can be strengthened by making them trumpet-shaped where they join the relatively thicker shaft (Figs. 1.272b and d). The shape shown in d prevents the lubricating oil escaping from the bearing, since the oil cannot creep beyond the major diameter of the journal. With a barrel-shaped shaft in a cylindrical bearing (Fig. 1.272e), the lubricating oil is drawn to the position of least clearance.

Fixed shafts can be incorporated in the structure by forcing or driving in, screwing and riveting (Figs. 1.273a, b and c). A lightweight rotor can be supported by two bearing jewels on a taut wire.

Other possible methods that have been adopted, particularly in mass production, are illustrated in Figs. 1.273d, e, f, g and h. In d and e, the shaft itself is used as a punch, and remains clamped in position as a result of the elastic deformation that occurs during punching. In f, the shaft is inserted in a wider hole, after which the hole wall is closed-up by staking.

Fig. 1.271. Frictional torque exerted on journal in bearing.

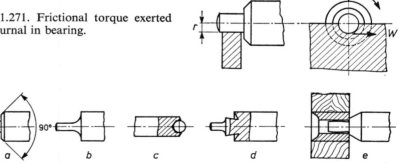

Fig. 1.272. Shaft journals.

    a. Journal turned on shaft.
    b. Turned, trumpet-shaped journal, for diameters less than 1 mm.
    c. Ball riveted in.
    d. Turned journal with oil trap.
    e. Barrel-shaped shaft in cylindrical bearing.

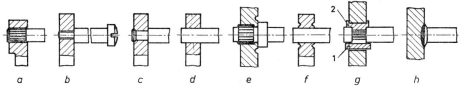

Fig. 1.273. Methods of fastening fixed shafts.

a. Knurled and driven in.
b. Screwed in.
c. Riveted.
d. Driven into hole punched by shaft itself.
e. As d, but locked.

f. Staked in clearance hole.
g. Clamped by deformation of extra ring.
  1 = ring before deformation.
  2 = ring after deformation.
h. Butt-welded.

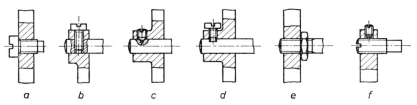

Fig. 1.274. Methods of screw-fastening fixed shafts.

a. Journal is part of cheese-head screw.
b. Shaft secured by cheese-head screw.
c. Shaft fixed by grub screw in dimple.
d. Shaft fixed by grub screw in milled slot.
e. Shaft screwed-in and locked with nut.
f. Journal on grub screw, locked by another grub screw with a soft copper plug under it to prevent damage to thread.

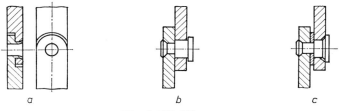

Fig. 1.275. Hinge.

a. Riveted journal integral with one of two members.
b. Riveted shaft with shoulder to ensure adequate clearance.
c. As b, but with extra washer to reduce frictional torque. Countersunk head reduces depth of assembly.

To produce a slightly heavier structure, a ring of very ductile material (e.g. aluminium) is staked between the shaft and the wall of the hole, as at g. The shaft in h is fixed by butt-welding. Journals are sometimes screwed-in or otherwise fastened by screws, but, a point to remember about this method is, that the journal cannot be centred exactly in its hole (Fig. 1.274).

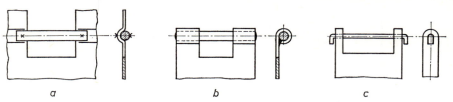

Fig. 1.276. Methods of fastening hinge-pins to hinges made of sheet material.

a. Spot welded pin.   *b.* Soldered pin.   *c.* Pin with ends bent over.

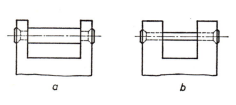

Fig. 1.277. Fastening the hinge-pins of hinges made of cast material, for example.

a. Pin with turned journals fastened by riveting.

b. Smooth, riveted hinge-pin with hinge robust enough to resist deformation during riveting.

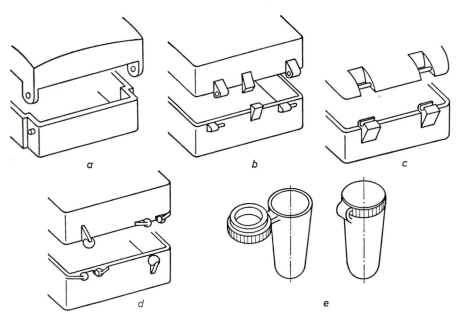

Fig. 1.278. Plastics hinge construction.

a. Lid fitted at elevated temperature, so that the two flanges can bend.
b. Lid can be fitted or detached in a specific position.
c. Box made of elastic material, so that undercut sockets are nevertheless detachable.
d. Hinge consisting of two round members rotating under initial stress in cupped sockets: lid and box are identical.
e. Cap and container of flexible material (polyethylene).

Hinges are best kept as simple as possible. Figs. 1.275 to 1.280 inclusive show structures of metal and of plastics. A polypropylene hinge for integration in the moulding (Fig. 1.279) closes without squeaking, if it is properly designed. The direction of flow (1) is set at right-angles to the bending direction, and steps are taken to ensure that the hinge does not contain a yield line. Polypropylene can withstand bending a great many times, particularly when it has a very smooth surface. A bead (2) along the edge of the hinge checks any tendency to split.

Three detachable piano hinges of different types are shown in Fig. 1.280.

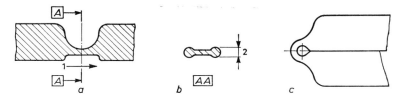

Fig. 1.279. Shaping of polypropylene hinge.

a and *b*. Cross-sections of the flexible hinge strip.
c. Closed hinge.
1 = direction of flow of material.   2 = thickened edges.

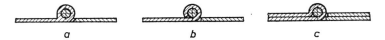

Fig. 1.280. Piano hinges.
a. Drilled hinge for heavy structures.
b. Rolled hinge for lightweight structures.
c. Pressed hinge, the most common type.

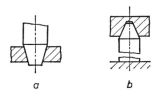

Fig. 1.281. Conical journal.

a. Journal rotating in fixed bearing
b. Fixed journal as support.

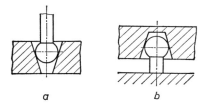

Fig. 1.282. Spherical journal.

a. Rotating journal in fixed, conical bearing.
b. Fixed journal as support for a conical bearing.

1.5.3 *Conical and spherical journals and bearings*[13–18, 34]

Conical journals are used (amongst other things) in geodetic instruments and in telescopes. A conical journal can also be employed as a fixed support: as such, it is not affected adversely by wear and is devoid of radial backlash (Fig. 1.281).

A spherical bearing can be set in different positions. The frictional torque is the same as that of a taper roller bearing having the same cone vertex angle. Because the contact surface of the sphere is small, spherical journals are only suitable for relatively light loads (Fig. 1.282).

For greater efficiency a conical journal should be constricted centrally, so that it does not bear over its full length. The actual supporting surface of the bearing should then extend slightly beyond the corresponding surfaces of the bearing in the direction of force, as shown in Fig. 1.283. A vertex

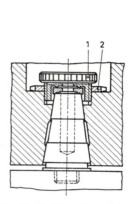

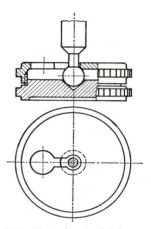

Fig. 1.283. Conical bearing with adjustable axial strain.
1 = grub screw.
2 = lock-nut.

Fig. 1.284. Clamping ball joint used on camera stands.

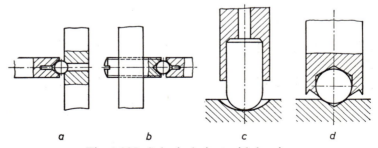

| a | b | c | d |

Fig. 1.285. Spherical pivot with bearing.

a. with loose ball;   *c.* with driven pin with spherical end;
b. as *a* and adjustable with grub screw;   *d.* with embedded ball.

angle commonly used is 7·5°, which is associated with a considerable amount of friction; this can be reduced by supplying an adjusting screw, to relieve the axial strain on the shaft.

Clamping ball joints for quick assembly are sometimes used on camera stands (Fig. 1.284). Figure 1.285 shows types of spherical pivot with (adjustable) bearings in actual use. The pivot shown in Fig. 1.285c has a minimum of friction when mounted vertically, since the frictional torque arm is short; theoretically, in fact, it should be non-existent. The ball in Fig. 1.285a is sometimes magnetized to prevent it from dropping out. Inexpensive assemblies of this kind can be made with balls.

The small-diameter journal for use where a free-running bearing is required has already been mentioned in Section 1.5.2. Frictional torque must be kept very small indeed in, for instance, moving-iron instruments, so a spherical, vertically-mounted pivot is preferred here. With the shaft or spindle horizontal (Fig. 1.286), the (small) clearance (1) required prevents any coincidence of the centre-lines of the revolving shaft and the stationary cup or seat. This is a normal state of a pivot. The pointed journal of the shaft is rounded with a radius r and merges into a conical shape. Contact between journal and seat occurs at the point marked A; the frictional torque is the product of frictional force $W = \mu N$ and arm r_1, where μ is the coefficient of friction. N is normal to the tangent plane at the point of contact, so it follows from the diagrams that:

$$N = \frac{G}{\sin \alpha} \quad \text{and} \quad \sin \alpha = \frac{r_1}{r}$$

hence

$$N = \frac{G \cdot r}{r_1}$$

The frictional torque $\mu N \cdot r_1$ is therefore equal to $\mu G \cdot r$. This demonstrates that the frictional torque of a pivot with a horizontal axis is proportional to the coefficient of friction, the weight of the rotor and the radius of the rounded end.

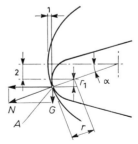

Fig. 1.286. Dimensions of a pivot and forces acting thereon.

2 = clearance.
1 = eccentricity of journal relative to cup.

This radius may be very much smaller than the smallest radius possible for a cylindrical journal. A pivot therefore reduces frictional torque to a minimum.

According to Hertz, the specific pressure at the point of contact is very high, much higher than in a plain bearing. Thus, pivots have to be made of very hard material, such as hardened steel, corundum (sapphire and ruby) and spinel. See also Volume 1, Section 2.3.13, and Volume 3, Section 4.2.

Impact on a pivot may destroy the bearing point, already under a heavy specific load. Figure 1.287 shows a cup protected against impact from different directions. This structure ensures correct location of the journal. Although it would be reasonable to expect only a small amount of frictional torque from a vertical pivoted shaft, the tilt effect (illustrated in Fig. 1.288) may give rise to frictional torque comparable to that set up by horizontal shafts. Many practical versions of pivots for horizontal shafts are dimensioned as follows (Fig. 1.289):

- the apex angle of the cone of the cup is 90°;
- the apex angle of the conical pivot, or journal, is 60°;
- the ratio of cup radius to journal radius is $R/r = 1.5$ to 10.

Pivots subjected to shock loads are given a ratio $R/r < 2.5$ and little axial clearance. A ratio $R/r > 3.5$ is preferred for bearings in ordinary measuring instruments, since the transition from the spherical to the conical surface of the cup then results in only a minor change in the specific pressure of the journal on the cup. Depending on the weight of the shaft with its associated parts, and on the permissible specific loads, journal radii r ranging from 5 μm to 50 μm or more are chosen. The coefficient of friction is about 0.13 for a hardened steel journal on a sapphire cup.

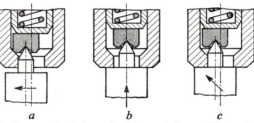

Fig. 1.287. Cup protected against impact from different directions. The structure ensures correct location of the journal.
- *a.* Situation after lateral impact.
- *b.* Situation after axial impact.
- *c.* Situation after impact from an oblique angle.

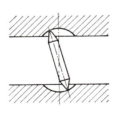

Fig. 1.288. Tilt effect associated with vertical-axis pivots.

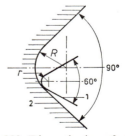

Fig. 1.289. Dimensioning of pivots. 1 = conical journal.   2 = cup.

1.5.4 Rolling-motion (anti-friction) bearings[19 − 25, 35, 36]

(a) Introduction

In anti-friction bearings (or rolling-motion bearings, as they are sometimes called) sliding motion is superseded by rolling. This greatly reduces the coefficient of friction, and makes (miniature) anti-friction bearings suitable for use in instruments. The slight frictional resistance also occurs on starting and does not vary much with the speed. Anti-friction bearings take up more room than plain bearings. The life of an anti-friction bearing is limited by fatigue effects, as a result of which the loading capacity is inversely proportional to the speed of rotation and the useful life required of them. Most bearings are made of stainless steel A.I.S.I. 440C (American designation): the hardness after heat treatment is around 59 Rockwell C and remains so up to 350°C. The German ball-bearing steel 100 Cr6 has a hardness after heat treatment of about 62 Rockwell C, which it retains up to 120 °C. For further particulars of ball-bearing steels, see Volume 2, Section 1.4.3.

Bearing manufacture imposes the most stringent requirements on accuracy and purity. The component parts are cleaned many times in special baths, and then assembled in workshops where dust concentration is kept to a minimum. After being inspected, the bearings are wrapped in waxed paper or packed in sealed polyethylene bags. Cleaning before installation is not only unnecessary, but harmful.

The overall dimensions and running accuracy of rolling-motion bearings have been standardized (Volume 1, Section 1.10). Miniature and instrument bearings are manufactured to the current series of sizes. Moreover, ranges of sizes are available, in which the bearing width is reduced to half the current size for the particular shaft diameter, and the outer race is of very much smaller diameter. Although the bearing can thus be accommodated in less space, the loading capacity is reduced (to 24%, for example, in the case of a 4 mm diameter shaft.

(b) Ball bearings

The principal types of ball bearing for fine mechanisms (Fig. 1.290) are:
  (i) Radial groove bearings (for mainly radial loading).
  (ii) Angular contact bearings (with shoulders).
  (iii) Axial groove (ball thrust) bearings.
  (iv) Pivot bearings and other special bearings.

(i) The radial groove ball bearing is suitable for many and various applications and is therefore the most common type. Several versions of it are made (Fig. 1.291):

● The ordinary kind.
● With flanged outer ring.
● With inner ring wider than the outer ring.
● With outer ring and inner ring offset slightly in the axial direction.
● With seals or shields.
● In different materials.
● In mm and inch sizes.

Flanges on a ball bearing facilititate installation, since a simple cylindrical hole, without retaining rings, is all that is required. The addition of a flange does not add to the width of the bearing.

Inner rings of greater width (projecting about 0·4 mm on both sides) facilitate construction, since retaining and packing rings can be dispensed with.

Bearings with outer and inner rings of equal width, but with the races so arranged that the rings are offset slightly in the axial direction, can be mounted in pairs under initial stress with simple tools. The bearings thus function as angular contact bearings in a structure without play (Duplex fitting).

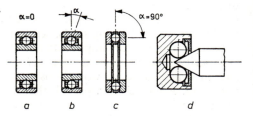

Fig. 1.290. Rolling-motion bearings for fine mechanisms.

  a. Radial groove ball bearing.
  b. Angular contact bearing.
  c. Ball thrust bearing.
  d. Pivot bearing.
  α = angle of contact.

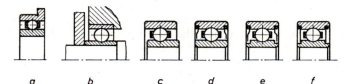

Fig. 1.291. Various types of rigid ball journal (radial groove) bearing.

a. Flange on outer ring.
b. Inner ring wider than outer ring.
c. Outer ring and inner ring offset.
d. Bearing with shields.

e. Bearing with shields and labyrinth sealing.
f. Bearing with seals.

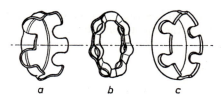

Fig. 1.292. Ball cage.

  a. Castellated cage of ball bearing steel.
  b. Two-part cage of steel strip.
  c. Snap-in type, made of synthetic resin-bonded fabric, or Delrin, sheet.

Bearings with shields or sealing rings have a neat appearance. The lubricant (grease or oil) has to be injected by the manufacturer. The shield or seal is fixed in a recess in the outer ring, and prevents the lubricant from escaping, at the same time excluding dirt. Closures with a small clearance between the running ring and the protective ring (shield), so that the two do not touch, are usual. Protective rings that bear slightly (seals) are not recommended, owing to the heat they generate, which is particularly important at high speeds. At the same time, sealing rings constitute a more effective closure than shields.

The following are used as ball cages for these bearings (Fig. 1.292):

Fig. 1.292a. A castellated cage with indented compartments for the balls, punched out of stainless, ball-bearing steel and hardened. This is the standard type of cage for low and average speeds.

Fig. 1.292b. A two-part cage made up of soft, ball-bearing steel strip; suitable for low speeds.

Fig. 1.292c. A snap-in cage made of laminated phenol sheet (synthetic resin-bonded fabric) or polyacetal sheet (Delrin). Machined and oil-impregnated laminated cages are primarily suitable for very high speeds at temperatures below 135 °C.

Machined Delrin cages (up to 143 °C) have little frictional resistance at low (or high) speeds and are exceptionally quiet. But, like metal cages, they cannot be impregnated.

Teflon distance-pieces are used to separate the balls in the bearings; they are of minimum width to ensure low frictional resistance at slow speeds.

(ii) The *angular contact bearing*, seen earlier in Fig. 1.290b, is used for relatively heavy and predominantly axial loads. An axial force can only be absorbed in one direction.

The bearing, the inner ring of which has a shoulder on one side only, is detachable. Less common is the (non-detachable) angular contact bearing with the shoulder on the outer ring. Towards larger contact angles, the permissible axial load increases and the axial resilience decreases. A cage of laminated material, impregnated with oil, makes high speeds possible. Angular contact bearings for Duplex installation are also obtainable.

(iii) The *axial groove*, or *ball thrust bearing* (Fig. 1.290c) is used exclusively to absorb (substantial) axial forces. The bore of the housing ring is slightly larger than nominal size, and the outer diameter of the shaft ring slightly smaller than nominal size, so as to avoid contact between the rotating and stationary parts without special modification to the structure. Such bearings are used, for example, in high-speed rotary liquid flow meters.

(iv) The *pivot ball bearing* (Fig. 1.290d) is a special type of angular contact bearing. It requires a very smooth, conical pivot (apex angle 60°) made of hardened (stainless) ball bearing steel: it can carry relatively heavy axial loads. When necessary, the bearing is prestressed to provide the axial force required for satisfactory performance (without play). The dished outer ring (Fig. 1.293) is made of high-grade strip steel.

Apart from the one above, the following other forms of pivot bearing are obtainable (Fig. 1.294):

- Bearings suitable for a shaft in the shape of a truncated cone, so that the balls run on a larger circumference of the shaft.
- Bearings fitted with a shaft-ring, so that the shaft itself need not be hardened.
- Bearings in which this shaft-ring bears against a shoulder on the shaft via a spring pressure washer, to cushion any impact more smoothly.
- Bearings suitable for a through shaft.

Fig. 1.293. Prestressed pivot bearings in a weighing machine.

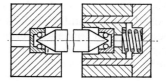

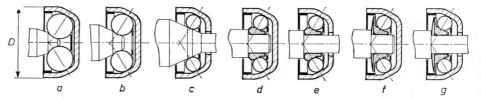

Fig. 1.294. Different types of pivot bearing.

 a. For conical journal ($D \geq 1 \cdot 1$ mm).
 b. For journal in the shape of a truncated cone ($D \geq 1 \cdot 1$ mm).
 c. As *b*, for through-shaft ($D \geq 2 \cdot 7$ mm).
 d. With shaft-ring ($D \geq 4 \cdot 25$ mm).
 e. As *d*, for through-shaft ($D \geq 4 \cdot 25$ mm).
 f. With shaft-ring and spring pressure washer ($D \geq 4 \cdot 25$ mm).
 g. As *f*, for through-shaft ($D \geq 4 \cdot 25$ mm).

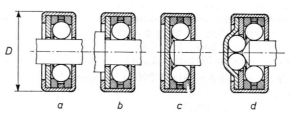

Fig. 1.295. Easy-running ball bearing for light load ($D \geq 3$ mm).

 a. For through-shaft with scope for linear motion.
 b. With polished plate to restrict axial motion of through-shaft.
 c. With polished disc as axial stop for end of shaft.
 d. With cover and balls to cushion axial thrust.

Easy-running ball bearings with a divided outer ring and without an inner ring, suitable for light reciprocating-motion loads, are shown in Fig. 1.295. The shaft bears directly on the balls and must therefore be made of hardened and polished ball bearing steel. If space permits, a bush or sleeve of hard material can be fitted on the shaft at the point of contact with the balls. Different types are used:

- For a through-shaft with scope for linear motion.
- With a polished plate to restrict axial motion of a through-shaft.
- With a polished disc as an axial stop for the end of a shaft.
- With a cover containing revolving balls to cushion axial thrust.

Fig. 1.296 shows three simple ball bearings constructed to cushion axial forces, and with members that can be integrated into the remainder of the structure.

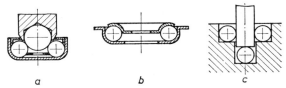

Fig. 1.296. Simple ball bearings constructed to cushion axial forces.
   a. Pivot with ball riveted into it.
   b. Detachable: can be integrated into remainder of structure.
   c. Combination of radial ball bearing and plain, axial bearing.

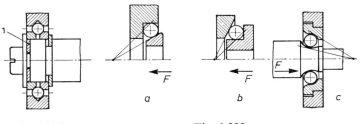

Fig. 1.297   Fig. 1.298

Fig. 1.297. Ball bearing with divided inner race: the clearance can be adjusted by matching the thickness of the washer (1).

Fig. 1.298. Ball bearings integrated in the structure.
   a. In pure rolling motion, the line through the two points of contact of the ball against the outer ring passes through the apex of the cone around the inner ring.
   b. In pure rolling motion, the line through the two points of contact of the ball against the inner ring passes through the apex of the cone in the outer ring.
   c. In pure rolling motion, the two tangents to the ball at the points of contact with shaft and outer ring intersect at a point on the centre-line of the bearing.

A version in which the inner ring is divided, and the outer ring is dispensed with (Fig. 1.297), fixes the shaft axially and gives scope for adjustment of play in the bearing.

Simple structures to cushion axial forces are shown in Fig. 1.298. Angular contact bearings integrated in the remainder of the structure are recommended for purposes of mass production, where the requirements for speed and running accuracy are not unduly stringent, and the price has to be kept low. It should be borne in mind that in this case the ball races are not produced in a factory specializing in them. They should be so designed as to ensure true rolling motion of the balls.

Another example of the special ball bearing is the *wire-seat ball bearing*, in which the ball seats are formed by four steel-wire rings set in grooves in the raceways. The advantages of this type of bearing are: accuracy, silence and small-space accommodation. In addition, the cost of large-diameter bearings is relatively low. The cross-section of such a bearing is shown in Fig. 1.299. The wire rings can be supported by any material, provided the shape of the structure ensures the necessary stiffness. In assembling the bearing, the wires are rolled-in by balls, which are afterwards replaced See also Fig. 1.300, of a turntable fitted with a radial wire-seat ball bearing

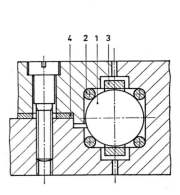

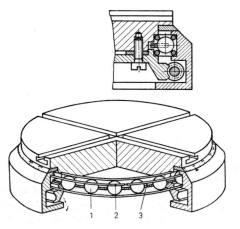

Fig. 1.299. Cross-section of wire-seat ballbearing.

1 = ball.
2 = steel wire ring.
3 = ball cage.
4 = packing ring governing play in bearing.

Fig. 1.300. Turntable supported by radial wire-seat ball bearing.

1 = ball.
2 = steel wire ring.
3 = ball cage.

(c) *Cylindrical roller bearings*

There are three main types of cylindrical roller bearing (Fig. 1.301), used for relatively heavier loads:

Non-detachable, to fix the shaft axially.

With detachable inner ring.

Without inner ring, fitted direct on the hardened and ground shaft.

The smallest of these bearings, for a shaft diameter of 4 mm, has an outside diameter of 12 mm. None of them, including the detachable type, can cushion axial forces.

Fig. 1.301. Cylindrical roller bearing.

   a. Non-detachable.
   b. Detachable inner ring.
   c. Without inner ring.

a   b   c

(d) *Fitting*

In principle, the factors governing the fitting of roller bearings are the same as for the larger ball journal bearings. Fitting should be done exactly in accordance with the manufacturer's instructions. The position of the shaft should be fixed precisely in relation to the housing, with due regard to the possibility of thermal expansion. If the load rotates relative to a ring, this must be mounted with an interference fit: if the load is stationary in relation to a ring, a sliding fit will do.

Other points to bear in mind are:

● Press fits must be avoided.

● The seating faces on the housing and the shaft should have the same accuracy of surface finish, and the same dimensional tolerance, as the corresponding faces on the bearing.

● The seating, or abutment, faces must be aligned accurately to within 1/4°.

● No force may be exerted on the rollers in mounting or demounting the bearings.

The use of adhesives to fix the bearing has certain advantages, and also certain drawbacks:

● Axial play in assembly is avoided.

● Press fits are unnecessary; the glue fills any reasonable clearance completely.

● Given a suitable choice of adhesive, e.g. Loctite, the running accuracy can be improved by revolving the shaft slowly whilst the adhesive is setting.

On the other hand:

● Some adhesives are attacked by lubricants or solvents.

● A sound adhesive joint can only be obtained if the (lubricated) bearing, the housing and the shaft are all perfectly free from grease.

● Damage may be caused by adhesive creeping into the bearing, owing to injudicious application.

● Vibration may cause the adhesive joint to give way.

In the absence of any special requirements, shafts are supported in the manner of Fig. 1.302. Tolerances and thermal expansion are taken into account. There are different structures which require *bearings without backlash* or bearings having a predetermined initial stress (pre-loaded bearings). This cannot be accomplished merely by taking bearings from an ordinary production batch and using them as they stand. However, the requirement is fairly easy to meet through axial readjustment of the bearing or by suitable pairing (Duplex fitting) (Figs. 1.303 and 1.304). Also, bearings can be fitted free from backlash and without initial stress by means of packing rings or nuts (Fig. 1.303).

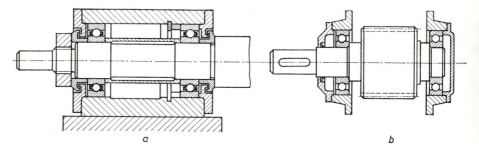

Fig. 1.302. Shaft in two bearings.

a. Fixed axially by right-hand bearing: outer ring of left-hand bearing can slide.
b. Outer rings of both bearings can slide; linear motion restricted by the two covers.

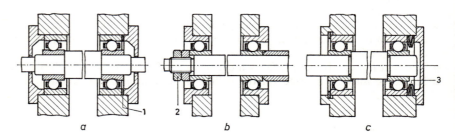

Fig. 1.303. Bearings without backlash.

a. Adjusted by means of packing ring (1).
b. Adjusted by means of nuts (2).
c. Thermal expansion compensated by dished spring washers (3).

Fitting with dished spring washers is recommended as a means of compensating thermal expansion (Fig. 1.303c). In this structure, the springs should supply sufficient thrust to keep the bearing in the correct position despite axial impact at rest. Unless the resilience of the springs is largely compensated by the force of the bearing in operation, the frictional resistance may easily be from 5 to 10 times greater than normal. Paired bearings, obtainable as radial groove ball bearings or as angular contact bearings, can be fitted in two ways (Fig. 1.304), namely, in what is called the O-arrangement, or in the more usual X-arrangement. If the bearings are pressed completely together on assembly, they are free from backlash or preloaded slightly, depending on the size of the original gap.

In practice, an initial stress or preload, equivalent to three times the force set up in operation, is adopted. Instead of pressing the bearings together, they can be mounted some distance apart by spacers or distance-pieces of equal length (Fig. 1.304b and d). It is advisable to calculate the axial spring action as a function of the bearing force.

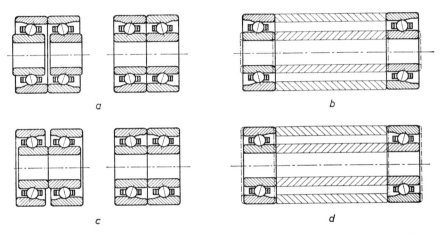

Fig. 1.304. Ball bearings rendered free from backlash by fitting them in pairs.

    a. O-arrangement (gap between inner rings): before and after fitting.

    b. O-arrangement: fitted some distance apart.

    c. X-arrangement (gap between outer rings): before and after fitting.

    d. X-arrangement: fitted some distance apart.

(e) *Loading capacity*

The size of the static load that can be imposed on instrument bearings is limited to a local compressive stress of 150×10^7 N/m². Heavier loads cause indentation of the raceway. Axial loading also involves the risk of damage in a shorter time, because of the ball pressing too closely against the edge of the raceway. A heavier dynamic load is acceptable, since any permanent deformation is distributed evenly around the raceway, so it does not affect the running of the bearing.

The following general rule has been adopted in the U.S.A. for determining the permissible dynamic load:

● For lightly loaded bearings, up to 5% of the dynamic load capacity.

● For average bearing loads, in excess of 15% of this capacity.

The relative permissible loads of bearings can be compared by determining the product Nd^2 of each bearing, where N = number of balls in the bearing and d = ball diameter. The loading capacity of bearings of similar design and material increases with this product.

The axial resilience of a bearing increases with the load. The resilience rises steeply at the outset, when the load is imposed, after which it gradually falls off. This explains the stiffness of preloaded bearings in that much of their axial resilience is already swallowed up in preloading. A substantial contact angle, and balls of large diameter, also add to the stiffness of the bearing.

(f) *Frictional resistance*

Instrument bearings are designed to have a minimum of frictional torque. This torque increases with:
- The dimensions of the bearing (frictional force acts on a longer arm).
- The load (more deformation).
- The speed (more resistance from the lubricant).

Lubrication is a critical factor. Generally, there is less friction in bearings lubricated with oil than in those lubricated with grease. But it may happen that greased bearings run with less friction than oiled bearings at high speeds, particularly when a grease is employed that remains in situ after the balls and the ball cage have channelled through it. The design of the bearing also affects the friction in it.

Light metal ball cages offer least resistance at low speeds, and plastic cages at high speeds. Frictional resistance in radial groove ball bearings and angular contact bearings is lowest at a contact angle around 15°. Small contact angles add to the resistance, because of geometrical errors in the ball races. Wide contact angles have the same effect, since the balls not only roll, but are compelled to slide. Specification of the coefficient of friction (μ) is sufficient for a rough estimate of the frictional torque, since this depends very much on the load. For radial groove bearings, an average $\mu = 0 \cdot 0015$ can be expected, referred to the radius of the shaft.

(g) *Speed of rotation*

The permissible speed for an instrument bearing is difficult to establish exactly, but it is possible to define the effect of a number of parameters on the possible speed, as follows:
- The highest speeds can be attained with small bearings, within the limits imposed by the (weak) ball cage.
- The permissible speed depends very largely on the load. Preloaded bearings restrict the speed by a factor of 5 as compared with a (similar) single-mounted bearing.
- Rotation of the inner ring permits a higher speed than rotation of the outer ring.
- Self-lubricating ball cages and mineral lubricating oils have to be employed at ultra-high speeds. Oil-mist and oil-injection lubrication have proved very successful here.

From the point of view of design, the following points are worth noting:
- An outer ring rotating at high speed requires a tighter fit, in view of the centrifugal forces acting on the housing and the ring.
- More bearing clearance is necessary to allow for possible thermal expansion in operation.
- The running accuracy and dynamic balance have to satisfy very stringent requirements.
- The same applies to the alignment and accuracy of the fitted parts.

In practice, the term "high speed" is applied to radial groove ball bearings and angular contact ball bearings for which the product of $n.d_m$ is greater than $0.4.10^6$, where n is the number of revolutions per minute and d_m the average bearing diameter (mm).

Grinding spindles, dental drills and centrifuges run at speeds up to the limit $n.d_m = 2.10^6$ of current engineering capability. For $d_m = 8$ mm, this is equivalent to a speed of 250,000 r.p.m.

(h) Noise[39]

Rolling motion bearings that are relatively silent-running are required for domestic appliances, office machines, medical apparatus, tape recorders and fractional horsepower motors. Factors affecting the noise level are: the accuracy of the bearing members, the size of the clearance involved, the accuracy of form and alignment of the housing and the speed of rotation. Through the choice of form and material, the structural design of the housing can contribute substantially to the damping of the noise generated, and to its prevention. Damping can also be improved by grease.

1.5.5 Methods of retaining shafts and bearings[38]

In principle, shafts can be retained in many different ways (Fig. 1.305). Collars are bulky and expensive to use. So circlips (sometimes called snap rings), punched or made of wire, are used instead. They permit quicker assembly and release, have smaller overall dimensions and are simpler. The hardened spring rings are snapped into mating grooves, which must be made to exact specification. Circlips also offer distinct advantages, compared with dowels, spring dowels, fluted pins and split pins. Fig. 1.306 shows round-wire circlips, circlips with a two-way spring action and triangular circlips. The first of the these circlips, shown at a, takes up very little room but requires a deep groove, equal to exactly half the thickness of the wire. If the groove is too deep, the members are not retained and slide over it; if the groove is too shallow, the clip slips out. If the wire used is too thick, the clip is permanently deformed as it is pushed on to the shaft. The use of circlips is thus limited to small-diameter shafts.

With the other two circlips shown, the full thickness of the wire drops into the groove, but they are only suitable for relatively light loads. Punched circlips (retaining rings) predominate. Their object is to dimension the radial section so that the stress set up when the ring is clipped on remains just below the yield point of the material throughout the ring, in order to get maximum flexibility without deformation.

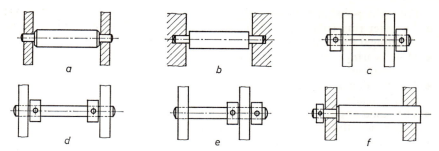

Fig. 1.305. Axial retention of a shaft supported in two separate bearings.

    a. Thicker part of shaft between the two walls.
    b. Journal ends confined in blind holes.
    c. Collar on each end of shaft outside the two walls.
    d. Two collars on shaft inside the walls.
    e. Two collars on shaft, one on each side of a wall.
    f. Thicker part of shaft on one side, and thinner journal with collar
        on other side, of one wall.

The groove is deeper than the thickness of the hardened spring clip, which is harder than the material grooved. One or two common types of retaining ring, for axial assembly, are shown in Fig. 1.307.

Some of them are bowed slightly to cushion axial backlash in the bearing. The E-spring, as it is called, shown in Fig. 1.308 is assembled radially, whilst the split, self-clamping circlip in Fig. 1.309 is for fitting on a smooth shaft: it can cushion moderate forces in both directions. Because there is no groove in the shaft, this type also provides compensation for backlash. Circlips of the kind shown in Fig. 1.310 are no longer used, because they are subject to plastic deformation and are extremely difficult to remove when once assembled.

It is necessary (or at any rate advisable) to use special tools for fitting and removing all these retaining rings. Specially-designed locking plates for the axial retention of shafts are shown in Fig. 1.311.

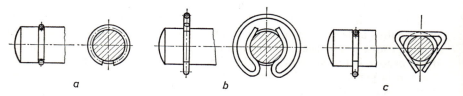

Fig. 1.306. Spring wire retaining rings.

a. Round wire ring in a groove of depth equal to half the wire thickness: for axial assembly.
b. Circlip with double spring action for radial assembly in a deep groove.
c. Triangular clip, used only for shaft diameters $\geq$ 10 mm: for axial assembly in a deep groove.

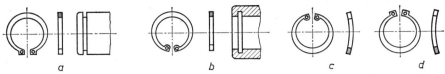

Fig. 1.307. Stamped retaining rings for axial assembly on a shaft or in a housing.

    a. For shafts: external type.

    b. For housings: internal type.

    c. As *b*, but bowed to cushion axial backlash elastically.

    d. As *a*, but bowed to cushion axial backlash elastically.

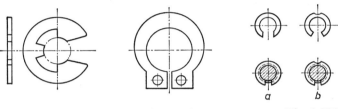

Fig. 1.308   Fig. 1.309   Fig. 1.310

Fig. 1.308. E-spring or circlip: for radial assembly.

Fig. 1.309. Self-clamping circlip (retaining ring) for axial assembly on a smooth shaft.

Fig. 1.310. Circlips (compression rings). Ring *b* deforms slightly more readily than ring *a*. No longer used because they are subject to plastic deformation and are extremely difficult to remove.

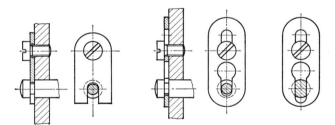

Fig. 1.311. Retention of shafts by different types of locking plate.

One or two points to bear in mind concerning structures involving anti-friction (rolling motion) bearings (Figs. 1.312 and 1.313) are:

- That axial forces must be cushioned by the flat of the raceway ring. A retaining ring must project beyond the shaft or housing enough to enable it to rest at an adequate distance from the radius of the ring (Figs. 1.312*a* and 1.313*a*).

- That placing a packing ring (2) between the bearing ring and the retaining ring (Figs. 1.312*b* and 1.313*b*) enables the bearing to transmit much greater axial forces.

- That it should be possible to demount the bearing by withdrawing a retaining ring, without exerting any force via the rolling members. This puts a certain amount of strain on the shoulder, on a shaft, or rim, in a housing (Figs. 1.312c and d and 1.313c and e).
- That the transition from the journal of the shaft to the level of the shoulder should be rounded with a radius smaller than the radius on the inner ring (Fig. 1.312c). Very much the same applies in the case of the rim in the inner housing and the outer ring. Recesses are recommended where necessary for the grinding operation (Fig. 1.312d), but must not weaken the shaft unduly.
- That the recommendation of the ball or roller bearing manufacturer should be followed exactly, so as not to prejudice the running accuracy and life of the bearing.

Very simple bearings, having a relatively short life, are found, for example, in switches. Figure 1.314 shows three cheap examples with solid spindles, and spindles made of tube and of rolled strip. The spindles are retained axially in a ceramic bearing.

Figure 1.315 illustrates a simple method of eliminating backlash from the bearing, with axial retention of the spindle by means of a wire spring.

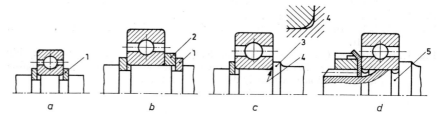

Fig. 1.312. Assembly of ball bearing on shaft.

a. Between two retaining rings (1).
b. As a: packing ring (2) can cushion greater forces.
c. Between retaining ring and shoulder (4).
d. Backlash-free assembly by means of nut and retaining ring.
1 = retaining ring.   2 = packing ring.   3 = shoulder.
4 = radius, to leave enough room for inner ring.   5 = recess.

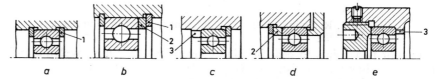

Fig. 1.313. Assembly of ball bearing in housing.

a. Between two retaining rings (1).
b. As a: extra packing ring (2) can cushion greater forces.
c. Between retaining ring and rim.
d. Assembly without backlash by means of ring screwed on.
e. Assembly without backlash by means of locked ring.
1 = retaining ring.   2 = packing ring.   3 = demounting slot.

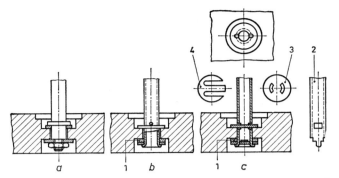

Fig. 1.314. Very simple bearing in ceramic housing of rotary switch.

a. Solid spindle with axial retention, subject to backlash owing to thickness of assembled rings: suitable for job production.

b. Tubular spindle with axial retention, subject to backlash dictated by split pin and flange: suitable for batch production.

c. Spindle of rolled strip, with axial retention, with backlash governed by two special rings in matching slots in the spindle; suitable for large-scale batch production.

1 = paper washer to protect the housing during rotary riveting.

2 = rolled spindle before assembly of *c*.

3 and 4 = special retaining rings.

The construction of the bearing, free from axial and radial backlash of the worm shaft shown in Fig. 1.316 is much more complex, but does not call for very accurate dimensional tolerancing.

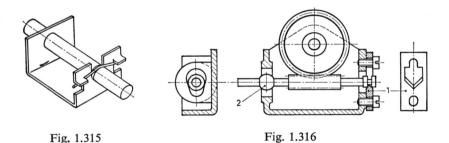

Fig. 1.315   Fig. 1.316

Fig. 1.315. Construction of a bearing eliminating the backlash from the bearing with axial retention of the spindle by means of a wire spring.

Fig. 1.316. Worm transport when the bearing of the worm shaft is free from axial and radial backlash and the clearance of the teeth can be adjusted by means of bearing plane (1). This V-shaped bearing construction eliminates the clearance in combination with the spherical shaft end (2) fixed in a V-shaped bearing recessed in the housing.

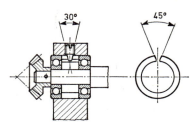

Fig. 1.317. Simple and efficient construction of a shaft fixed in two radial groove ball bearings.

A mounting example of a short shaft fixed in two bearings is given in Fig. 1.317. A grub screw with point fixes the total construction by intervention of a split ring by which the position of the bearing can be adjusted axially.

1.5.6 Special bearings[34]

(a) Knife-edge bearings[26] [40]

For these knife-edge bearings the movement is largely a rolling one. The greater the radius of the bearing seat, the smaller the sharp edges (knives), the smaller the oscillating angle and the more the movement approaches pure rolling. This kind of bearing is applied as supporting bearings in weighing machines and balances. The knives are made from hardened steel or jewels. Jewel knives, applied in rooms where acid vapour is much in evidence, may only be loaded up to 2 N. The knife has a prismatical shape and can have a triangular, olive or square cross-section. The angle of the knife for steel is 45 to 90°, and for jewels is 90 to 120°.

An obtuse angle is required for heavy loads. The sensitivity of the bearing increases with the acuteness of the angle as also the specific load, so that a knife-edge bearing has to be shielded against impact. Knife-edges in scales are invariably attached to the levers, because this enables the correct arm length to be obtained. The knife-edge is pressed into the carrier (steel knife-edges) or held in this by screws (jewel knife-edges), which usually give some scope for adjustment (Figs. 1.318 and 1.319). As a rule, the bearing seat is likewise vee-shaped.

Figure 1.320 shows the working parts of a knife-edge bearing with a vee-shaped seat in cross-section and also gives the dimensions commonly employed. Round bearing seats are easier to grind. Flat seats, or steps, are also used in very accurate scales. They are made of very hard steel or jewels (sapphire or ruby). Seats are fixed in much the same way as knife-edges, but if there is any fear of damage as a result of pressing in, an epoxy adhesive is used instead.

Knife-edge bearings are employed not only in scales and balances, but in measuring instruments and relays, and for other purposes involving only a small deflection and permitting only a small amount of friction (Fig. 1.321). Figure 1.322 shows a microprobe with three knife-edge bearings.

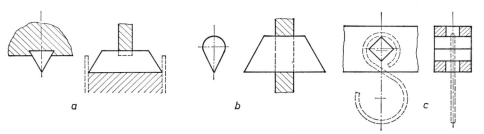

Fig. 1.318. Knife-edges pressed or glued in matching slots.

a. Knife-edge of triangular cross-section.   c. Knife-edge of square cross-section.
b. Knife-edge of pear-shaped cross-section.

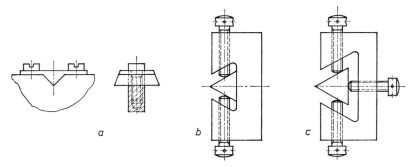

Fig. 1.319. Knife-edges, screwed into the carrier, are adjustable.

a. With two cheese-head screws.   c. With six set-screws: gives scope
b. With four set-screws.   for vertical adjustment.

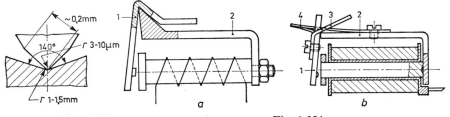

Fig. 1.320   Fig. 1.321

Fig. 1.320. Cross-section of the working parts of a knife-edge bearing with a
           vee-shaped seat, as used in scales.
Fig. 1.321. Knife-edge bearing in a relay.
           a. When the magnet picks up, the armature (1) pivots through a narrow
              angle on the knife-edged yoke (2). The shape of the yoke presents
              problems in manufacture.
           b. Improved version of a. The armature takes the form of a flat plate (1)
              backed-up by the bent armature lever (3). The two together form a
              vee-shaped bearing seat which is held on the yoke (2) by a spring (4),
              whose tab, together with plate and armature lever, constitutes
              another knife-edge bearing.

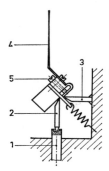

Fig. 1.322. Microprobe with three-knife-edge bearings.
  1 = probe.
  2 = movable knife-edge.
  3 = fixed knife-edge.
  4 = pointer.
  5 = adjusting screw.

(b) *Magnetic bearings*[27, 28]

These bearings are based on the mutual repulsion or attraction of built-in magnets. Some of them are magnetically load-relieved; others carry the full weight of the rotor. The object in both cases is to obtain very low values of frictional torque, and therefore wear.

Figure 1.323 shows the load-relieved thrust bearing of a 140-gm rotor in an electricity-supply meter, the pivot of which carries only one-sixth the full weight. Figure 1.324a shows a magnetic hanger bearing and Fig. 1.324b a magnetic thrust bearing. Magnets carry the full weight of the rotor in both of them. Because the equilibrium is not stable in all directions, any (minor) transverse forces have to be cushioned by extra bearings at both ends of the rotors. Bearings of this kind are used in American a.c. meters, in which the rotor weighs only 30 to 40 gm. The magnets in them must be capable of withstanding leakage flux. The weight of the magnet as mounted in the rotor must be small.

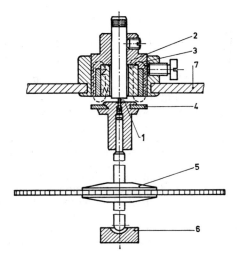

Fig. 1.323. Magnetically load-relieved thrust bearing of a rotor.
  1 = needle bearing to cushion trans-
      verse forces.
  2 = annular permanent magnet.
  3 = magnetic circuit.
  4 = round disc of soft iron.
  5 = rotor.
  6 = load-relieved bearing.
  7 = body of motor,

Fig. 1.324. Magnetic wire-bearing for rotor.

a. Hanger bearing.
b. Thrust bearing.
1 and 2 = permanent magnets.
3 = rotor.
4 = needle collar bearing to absorb lateral forces.

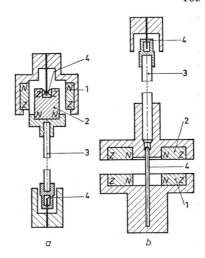

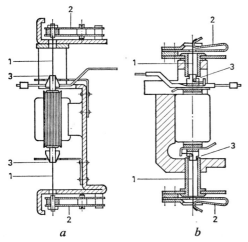

Fig. 1.325. Measuring frame with span-strip bearing.

a. Constructed with long strips (1) (about 27 mm) tensioned by adjustable parallel springs (2).
b. Constructed with short strips (1) (about 10 mm) to reduce overall height. The strip-ends are anchored outside the torsion-loaded system. The strips are tensioned by flat, hairpin springs (2). 3 = impact stop.

(c) *Span-strip bearings*[30]

These bearings are frictionless, and only suitable for rotation through limited angles. Such rotation involves the twisting of two thin, flat springs (strips), with the rotor shaft clamped between them, which supply a controlling couple. Any rotor current required is delivered via the strips.

In sensitive measuring instruments, the strips are not more than 0·1 mm wide, 0·01 mm thick and are made of bronze or an alloy of platinum and nickel. Impact stops prevent excessive lateral deviation (Fig. 1.325).

For particulars of *hinge springs*, see Section 1.7.5.

(d) *Air bearings*[29, 31]

In air bearings, the bearing faces are separated completely by a thin cushion of air. Since it is rarely possible to set up sufficient pressure in the bearing dynamically, compressed air has to be injected (aerostatic bearing). Because there is no metallic contact between shaft and bearing, there is no wear. The running accuracy can be greater than would otherwise be consistent with the local irregularities of the bearing surfaces. The friction is purely viscous and therefore increases with the speed. Because the viscosity of air is very low, however, friction, even at high speeds, is minimal. At low speeds, friction is extremely low and the bearing is completely free from slip-stick. The material, temperature and surroundings of the bearing are virtually unimportant.

Figure 1.326a shows the variation of air pressure under a shaft. This pressure is highest at the inlet and falls off gradually in both directions to atmospheric pressure. The smaller a gap, the more difficult it is for the injected air to escape, and the heavier the load the bearing can carry. The bearing clearance is not a true void, but is filled with a cushion of air, in such a way as to preserve a constant balance between load and deviation (Fig. 1.326b). Figure 1.327 shows a simple version of a radial air bearing.

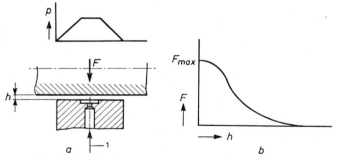

Fig. 1.326. Shaft or plate supported by a cushion of air (air bearing).

a. Pressure variation between inlet and outer edge of the bearing.
b. Continuous relationship between load and size of air-gap, that is the thickness of the supporting air cushion.

$$1 = \text{air inlet.} \qquad p = \text{air pressure under the shaft.}$$
$$F = \text{load.} \qquad h = \text{size of the gap.}$$

The accuracy of the motion and the absence of stick-slip have fostered the use of such bearings in measuring machines and machine tools. Because of its suitability for high-speed rotation, the air bearing has been adopted for grinding spindles (internal grinding), gyroscopes and dental equipment. Other prospective applications are in machines for the food industry, in medical apparatus and instruments, and generally where the action of oil and grease has to be avoided.

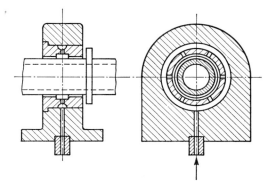

Fig. 1.327. Radial (aerostatic) air bearing with compressed air supply.

REFERENCES

[1] RICHTER, V. VOSS and KOZER, *Bauelemente der Feinmechanik*, Verlag Technik, Berlin 1954.
[2] K. HAIN, *Die Feinwerktechnik*, Fachbuch Verlag Dr. Pfanneberg, Giessen, 1953.
[3] G. SCHLEE, *Feinmechanische Bauteile*, Verlag Konrad Wittwer, Stuttgart, 1950.
[4] D. C. GREENWOOD, *Product Engineering Design Manual*, McGraw-Hill, New York/Toronto/London, 1959.
[5] K. RABE, *Grundlagen Feinmechanischer Konstruktionen*, Anton Ziemsen Verlag, Wittenberg/Lutherstadt, 1942.
[6] H. EDER and W. P. UHDEN, *Taschenbuch der Feinwerktechnik*, C. F. Winter'sche Verlaghandlung, Prien, 1965.
[7] K. RABE, *Lagerungen in feinwerktechnischen Geräten*, Feinwerktechnik, 1958, 6, p. 191.
[8] *Bearings Reference Issue*, Machine Design, 1966, March 10.
[9] F. WOLF, *Lagerungen, Geradführungen und Kupplungen in der Feinwerktechnik*, Deutscher Fachzeitschriften- und Fachbuchverlag, Stuttgart, 1956.
[10] A. KUHLENKAMP, *Konstruieren und Entwerfen in der Feinwerktechnik*, Konstruktion, 1963.
[11] G. REUTER, *Dichtungen und Lagerungen aus thermoplastisch verarbeitbaren elastischen Polyurethanen*, Kunststoffe, 1964, p. 530.
[12] J. L. VERMILLION, *How to Make the Polypropylene Hinge*, Modern Plastics, 1962, July.
[13] A. G. ASAFF, *Design it with Diamonds*, Product Engineering, 1961, October.
[14] Katalogus Seitz und Co, Les Brenets, Switzerland, Führer für die Verwendung synthetischer Steine.
[15] J. BUBERT, *Unsicherheit bei horizontaler Spitzenlagerung in Theorie und Praxis*, Feinwerktechnik, 1960, 9, p. 323; 10, p. 361 and 12, p. 431.
[16] J. BUBERT, *Betrachtungen über den Keinath Gütefaktor, die Einstellsicherheit und den Reibungsfehler bei elektrischen Messgeräten*, Feinwerktechnik, 1961, 7, p. 235; 8, p. 296.
[17] J. BUBERT, *Betrachtungen über die vertikale Spitzenlagerung bei elektrischen Zeigermessgeräten*, Feinwerktechnik, 1962, 11, p. 393.
[18] G. F. TAGG, *Maximum Pressure between Pivot and Jewel in an Instrument*, Instrument Practice, 1957, p. 130.
[19] A. PALMGREN, *Grundlagen der Wälzlagertechnik*, Stuttgart, 1964.
[20] R. R. PIERSON, *Instrument Bearings*, Bearings Reference Issue, Machine Design, 1966, March 10.

[21] Catalogue Instrumenten Kugellager, Gebrüder Reinfurt, Würzburg.
[22] Catalogue Roulements Miniatures Biennes, Miniaturwälzlager A. G. Biel, Switzerland.
[23] L. SCHARD, *Wälzlager in der Feinwerktechnik*, Feinwerktechnik, 1968, 1, p. 1.
[24] K. JANISCH, *Kugelrollspindeln und Wälzführungen in gesteuerten Fertigungseinrichtungen*, Feinwerktechnik, 1966, 6, p. 257.
[25] *Lager für Kreiselläufer*, Konstruktion, 1953, 5, p. 166.
[26] *Schneiden, Achsen und Pfannen*, DIN 1921, September 1964.
[27] G. SCHERTEL, *Daumermagnete in der Zählertechnik*, Feinwerktechnik, 1967, 1.
[28] S. HILDEBRAND, *Zur Frage magnetisch-entlasteter Lager*, Feinwerktechnik, 1964, 9, p. 383.
[29] R. LEHMANN, A. WIEMER and H. ENDERT, *Luftgelagerte Bauelemente im Feingerätebau*, Feingerätetechnik, 1957, 7.
[30] S. HILDEBRAND, *Zur Berechnung von Torsionsbänder im Feingerätebau*, Feinwerktechnik 1957, 6, p. 191.
[31] P. L. HOLSTER, *Gaslagers met uitwendige drukbron*, Polytechnisch Tijdschrift W., 1967, p. 363 and 415.
[32] J. M. EASTMAN, *Designing for less Friction*, Product Engineering, 1952, May, p. 124.
[33] D. D. CARSWELL, *Plastic Bearings*, Bearings Reference Issue, Machine Design, 1966, March 10.
[34] K. H. BERG, *Konstruktionsbeispiele aus der Feinwerktechnik*, Konstruktion, 1956, 5, p. 186.
[35] Catalogue Instrumenten–Kugellager, Gebrüder Reinfurt, Würzburg.
[36] Catalogue Miniature and Instrument Precision Rolling Bearings, The Hoffmann Manufacturing Co., Chelmsford, Essex, England.
[37] Catalogue Prestincert, Enfield, Middlesex, England.
[38] N. N., *Wat is de meest geschikte geponste borgveer?* Polytechnisch Tijdschrift A, 1951, No. 11/12, p. 194a.
[39] G. LOHMANN, *Untersuchungen des Laufgeräusches von Wälzlagern*, Konstruktion, 1953, 2, p. 38.
[40] E. FISCHER, *Der Bewegungswiderstand in verschiedenen Arten von Schneidenlagern*, Feinwerktechniek, 1968, 5, p. 217.
[41] R. L. PETERS, *Design of integral Plastic Hinges by Nomograms*, Design News, 19 July 1967, p. 82.
[42] S. HILDEBRAND, *Feinmechanische Bauelemente*, 1968, Carl Hanser Verlag, Munich.

1.6 Detents, clamps, ratchet mechanisms, stop mechanisms, etc.[1–6, 11]

1.6.1 Introduction

The purpose of the devices now to be discussed is to restrain a structural element temporarily in a guide permitting sliding or rotary motion. This restraint may prevent the latent motion. In other words, the mechanism may completely suppress, or merely limit or impede the remaining degree of freedom. At rest, these mechanisms are in effect detachable fastenings. As to freedom, the possible constructions can be brought under six headings (Fig. 1.328). The mechanisms *a*, *b* and *c* in the top row of the diagram operate through a retaining element shaped to block or impede the motion, always subject to certain discrete constraints:

a. Detent: blocks motion in either direction.

b. Ratchet and pawl: blocks motion in one direction.

c. Stop mechanism: impedes motion in either direction.

The mechanisms *d, e* and *f* in the bottom row of the diagram operate through a frictional member whose force arrests or impedes the motion:

d. Clamp: arrests motion in either direction.

e. Friction ratchet: arrests motion in one direction.

f. Brake: impedes motion in either direction.

The retaining or frictional member, and the member subject to temporary restraint, can both be designed to rotate or to slide. Hence the six possible versions shown in Fig. 1.329.

Other devices designed to limit stroke or travel, namely stops, will be discussed towards the end of this chapter.

Fig. 1.328. Six groups into which mechanisms designed to arrest or impede a motion can be divided.

a. Locking device.

b. Ratchet mechanism.

c. Stop mechanism.

d. Clamp.

e. Friction ratchet.

f. Brake.

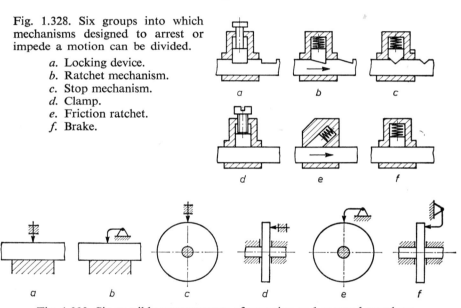

Fig. 1.329. Six possible arrangements of engaging and engaged members.

a. Both members slide.

b. Rotary member engages slide.

c. Sliding member engages rotary member radially.

d. Sliding member engages rotary member axially.

e. Both members rotate (radial engagement).

f. Both members rotate (axial engagement).

1.6.2 Locking mechanisms

A detent blocks motion in such a way as to preclude release by any force exerted on the member detained. A lead-in on the detent, for accurate location, etc., should be bevelled to an angle acute enough to prevent slipping. As explained in Section 1.4, it may happen that a sliding detent runs heavily or even sticks. This is not likely to happen with a rotary detent, however, because its frictional force acts on the periphery of a (small) journal, whilst in the event of sliding motion the friction is directly opposed to the applied force. Figs. 1.330 to 1.332 inclusive show one or two of these mechanisms.

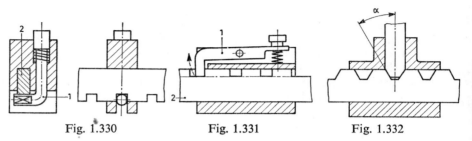

Fig. 1.330   Fig. 1.331   Fig. 1.332

Fig. 1.330. Sliding detent for a sliding rod. Detent is released by pressing a button.
           1 = detent.   2 = rod.

Fig. 1.331. Rotary detent engaging slotted rod requires some clearance in the slot in order to function. As a result, the rod cannot be fixed with any degree of accuracy.
           1 = detent.   2 = rod.

Fig. 1.332. Sliding detent for sliding rod. Lead-in on detent ensures accurate fixing. Angle α must be kept within given limits to make the device self-sticking.

1.6.3 Clamps

The clamping force stems from friction between the retaining and retained members. Substantial clamping forces are supplied by screws, keys or sometimes springs. Another convenient method is clamping by means of a short cylindrical or angular-section guide. There is a wide variety of possible designs, some of which are given in Figs. 1.333 to 1.340 inclusive. Figure 1.341 shows ways of restraining universal joints, whilst Figs. 1.342 and 1.343 illustrate the use of a clamping guide.

Fig. 1.333. Hub clamped on shaft by a screw.
a. With round-ended grub screw bearing on flat face of shaft.
b. With blunt-pointed grub screw engaging vee-groove in shaft.
These fastenings are not altogether free from backlash, because the shaft can twist slightly under the screw, depending on the clearance. Possible damage to the shaft does not prevent readjustment or release.

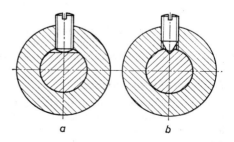

a   b

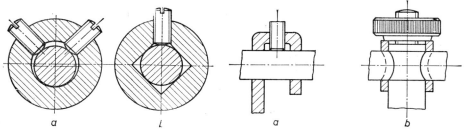

Fig. 1.334. Hub clamped on shaft in three-point fastening entirely free from backlash.

a. With two grub screws.
b. In square hole with one grub screw.

Fig. 1.335. Clamp fastening with three points of contact.

a. Lever on shaft with screw.
b. Nut draws rod tight, with aid of sleeve.

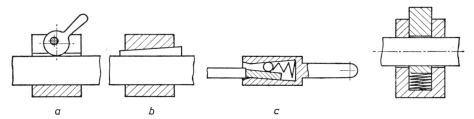

Fig. 1.336. Clamp fastening by wedge action with:
 a. Cam catch. *b.* Longitudinal key.
  c. Ball in conical recess.

Fig. 1.337. Clamping through friction set up in three places by spring.

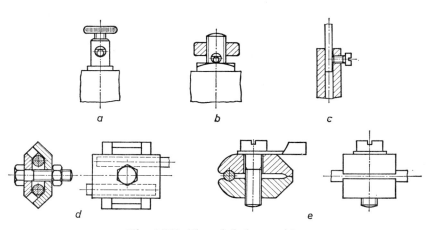

Fig. 1.338. Threaded clamp with:

a. Knurled screw. *c.* Cheese-head screw. *e.* Two jaws.
b. Knurled nut. *d.* Two clamps.

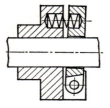

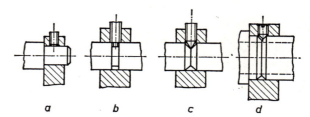

Fig. 1.339. Clamping
by short journal bear-
ing set askew by
spring pressure.

Fig. 1.340. Clamping a shaft.

a. Journal retained by grub screw: any damage to shaft
   impedes demounting.
b. Shaft retained by grub screw seated in groove;
   accuracy of axial retention depends on fit of screw
   in groove.
c. Accurate axial retention of shaft by pointed grub
   screw in vee-groove.
d. Shoulder of hollow shaft drawn against wall by one
   or more pointed grub screws seated in groove.

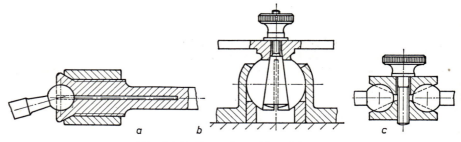

Fig. 1.341. Retention of ball joints.
a. Split bearing compressed by coupling nut.
b. Split ball journal opened out by key drawn into it.
c. Two clamps form split double bearing.

1.6.4 Ratchet mechanisms

They permit motion in one direction only. Undesired motion in the other
direction may be prevented by the shape of the structural members or by
a (frictional) force. Mechanisms operating on the latter principle are called
friction ratchets. Those of the former type contain an ordinary ratchet wheel
and pawl to restrain rotation, or a rack-and-pawl to restrain linear motion.
However, ratchets are used in almost all of them. The pawl may be loaded
in compression (push pawl) or in tension (pull pawl). Push pawls are
preferred, because it is easier to machine them to the correct alignment

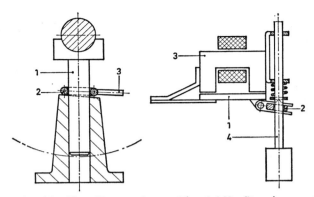

Fig. 1.342. Jack with pillar (1) passed through ring (2) which bears on an inclined face. Under load, the ring clamps against the pillar and can only be released by knocking the lug (3) (connected to the ring) upwards with a hammer.

Fig. 1.343. Stepping motor with an armature (1) hinged to a ring (2) having only a very small bearing face. When the a.c. magnet (3) is energized, the ring slides up and down. On the upstroke, the ring clamps against the rod (4), carrying this with it; on the down-stroke, the ring slides along the rod. Fast reciprocation draws the rod upwards.

(Fig. 1.344). Since the mechanism is subject to shock loading, the tip of the pawl should be rounded slightly to prevent it from sticking in the tooth space. To avoid undue surface pressure, the contact faces of tooth and pawl must mate exactly. Pawls are made of hardened and tempered steel to make them hard wearing. However, a softer material can be used for the ratchet wheel, the teeth of which are very much less prone to wear.

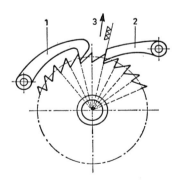

Fig. 1.344. Ratchet wheel with pawls: a push pawl is preferred.
1 = pull pawl, difficult to machine correctly.
2 = push pawl, readily machined to correct alignment (3).

Figure 1.345 shows the more important of the possible layouts for rotary ratchets and pawls. The location of the pawl pivot relative to the ratchet wheel is a critical factor (Fig. 1.346). Three basically different pivot locations for radial alignment of the working tooth face are shown:

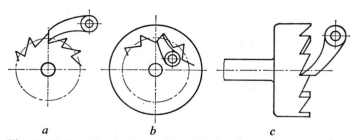

Fig. 1.345. Possible versions of pawl and ratchet wheel arrangement.
a. Ratchet wheel with external teeth.   *c*. Dished ratchet wheel.
b. Ratchet wheel with internal teeth.

- Pivot on the tangent to the pressure point on the ratchet wheel.
- Pivot inside this tangent (disengaging ratchet).
- Pivot outside this tangent (engaging ratchet).

In the last case, the pawl is drawn inwards, but allows the ratchet wheel to back slightly after it has dropped in. With the pivot inside the tangent, there is a resultant tending to force the pawl away from the ratchet wheel. The optimum pivot location is on the tangent to the ratchet, with, say, a spring to hold the pawl in contact with the teeth.

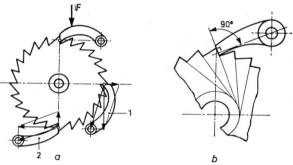

Fig. 1.346. Range of possible locations of pawl pivot relative to ratchet wheel.

    a. Location of pivot in relation to tangent to ratchet, with radial
        alignment of working tooth face.
    b. Working tooth face alignment suited to position of pawl pivot.
    1 = disengaging ratchet.   2 = engaging ratchet.   *F* = force of spring.

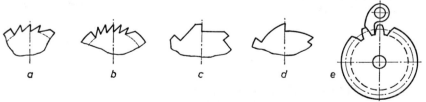

Fig. 1.347. Ratchet tooth shapes.

a, *b*, *c* and *d* have different pitches, but common radial alignment of working tooth face. *e*. Ordinary gear wheel as ratchet wheel.

Where this arrangement is impracticable, the tooth face is aligned so that the normal to the pressure point on it passes through the pawl pivot. This brings slightly greater forces to bear, which have to be cushioned by the bearings (Fig. 1.346*b*). Fig. 1.347 shows a number of different tooth shapes, depending on the number of teeth, the diameter of the ratchet wheel, the force acting on the tooth and the pawl position relative to the ratchet wheel. An ordinary gear wheel can also be used as a ratchet wheel, if the pawl is adapted to it.

If so many ratchet wheel settings are required that this would reduce the pitch of the teeth on a single ratchet wheel unduly, two identical wheels can be mounted on the same shaft, with the teeth on one of them staggered by half the pitch of those on the other wheel.

Figure 1.348 shows ratchet mechanisms with external teeth, a notable feature of which is the construction of the pawl pivot. Mechanisms with internal teeth and with several pawls to relieve the strain on the ratchet wheel are illustrated in Fig. 1.349, whilst Fig. 1.350 shows different springs used to hold the pawl in contact, and Fig. 1.351 illustrates the use of gravity for the same purpose.

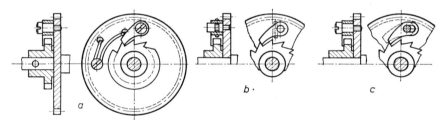

Fig. 1.348. Ratchet wheels with external teeth.

a. Shouldered screw as pawl pivot.
b. Shouldered screw with slot as pawl pivot.
c. Elongated hole in pawl drops round shouldered pin, thus permitting slight backlash of ratchet wheel, to prevent clockwork spring from being overloaded when wound too tight.

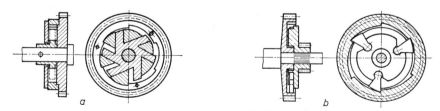

Fig. 1.349. Ratchet wheels with internal teeth.

a. With six triangular sliding pawls. Left-hand side of cross-section shows pawls engaged, right-hand side shows them disengaged.
b. With three rotary pawls.

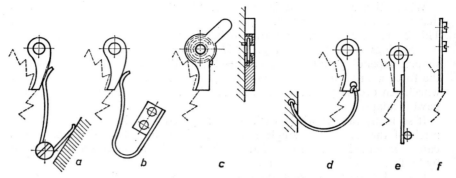

Fig. 1.350. Spring pawl-engaging systems with:

a. Wire hairpin spring.
b. Flat hairpin spring.
c. Circular wire spring; it takes up little room.
d. Semi-circular wire spring.
e. Flat spring fastened to pawl.
f. Flat spring fastened to frame.

Figure 1.352 shows a *friction ratchet mechanism.* Instead of being shaped to engage in a slot, the pawl has a smooth surface making frictional contact only, at point A. The ratchet wheel is free to rotate anticlockwise. To ensure fail-safe restraint of clockwise rotation, the angle must not be made too obtuse. The tangential friction W caused by clockwise rotation is μN, where N is the normal pressure between the pawl and ratchet wheel. The condition governing rotational balance of the ratchet wheel is:

$$\mu Nr \cos \alpha = Nr \sin \alpha \quad \text{or} \quad \sin \alpha = \mu \cos \alpha$$

and

$$\mu = \tan \alpha.$$

To restrain the ratchet wheel without fail, it is essential to make $\mu > \tan \alpha$.

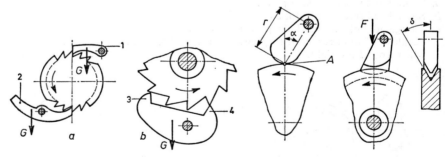

Fig. 1.351   Fig. 1.352   Fig. 1.353

Fig. 1.351. Pawl engages through action of gravity (G).
    a. Pawl (1) by own weight and pawl (2) by counterweight.
    b. Point (3) of pawl is pushed away from ratchet during rotation, leaving point (4) ready to drop in.
Fig. 1.352. Friction ratchet mechanism.
Fig. 1.353. Friction ratchet mechanism with wedge-shaped contact.

Fig. 1.354. Freewheel, a friction pawl mechanism with several balls to relieve the strain on the ratchet wheel.

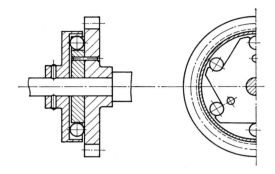

Taking $\mu = 0\cdot15$ gives $\alpha \approx 8° \, 30'$. The pressure on the pawl is $W/\sin \alpha \approx 7W$, which imposes very heavy loads on the bearing and also calls for accurate dimensioning, in order to preserve the correct angle α. A wedge-shaped contact between pawl and ratchet wheel is more satisfactory (Fig. 1.353), and increases the friction by a factor of $1/\sin \delta$, where δ is half the apex angle of the wedge. Thus $\mu\,.\,(N/\sin \delta)\,.\,r \cos \alpha > Nr \sin \alpha$, so that $\tan \alpha$ must be $< \mu/\sin \delta$. With $\delta = 15°$ and $\mu = 0\cdot15$, α may be $< 30°$ in the blocked state.

The *freewheel* (Fig. 1.354) is another friction pawl mechanism. The pawl may be provided with several balls spaced round the ratchet, to relieve the wheel spindle of forces set up by the pawl (see also Volume 1, Sections 2.3.9 and 2.3.10).

A special combination of several ratchet mechanisms operating with a common pawl is to be found in electronic appliances.

When any button of this *push-button mechanism*, as it is called, is pressed, that button is locked down by the common pawl until another button is pressed, releasing the pawl: the push-rods and the pawl can rotate or slide. The four possible versions are shown in Fig. 1.355, and in Volume 7, Chapter 2. Sliding rods are used where the movement need not change direction, but run more heavily than levers, which also enable the stroke of the movement to be lengthened or shortened.

A comparison of the sliding pawl with the rotary pawl reveals the following differences:

- The rotary pawl runs more lightly than the sliding pawl.
- The rotary pawl does not exert any lateral pressure on the push-rod, unlike the sliding pawl, which therefore calls for an effective rod guide to prevent it from running too heavily.
- The rotary pawl takes up more room than the sliding pawl.
- The sliding pawl constitutes a more difficult blanking process (hole profile) than the rotary pawl (Fig. 1.356).
- The positioning of the push rods is more critical in the case of the sliding pawl.

To ensure fail-safe release of a previously locked button when another button is pressed, a bevel is applied as shown in Fig. 1.356. Depending on the design the lever or the sliding pawl is bevelled. Thus, contact between pawl and push-rod/lever is made on an inclined face before the button to be pressed reaches the top of its stroke, thereby ensuring release.

Thus, the possible adverse effects are avoided, of:
- Inaccurate shaping of rod/lever and pawl.
- Pawl and/or rod/lever wear.
- Give on the part of the pawl and/or the remainder of the structure.
- Play in the guides or pivots of pawl or rod/lever.

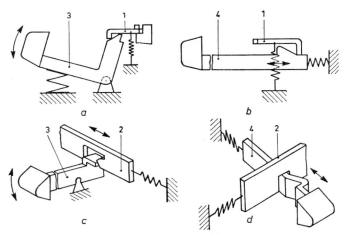

Fig. 1.355. Push-button mechanism for electrical appliances.

a. Push-button with lever and rotary pawl.   1 = common rotary pawl.
b. Push-button with slide rod and rotary pawl.   2 = common sliding pawl.
c. Push-button with lever and sliding pawl.   3 = lever with button.
d. Push-button with slide rod and sliding pawl.   4 = slide rod with button.

Fig. 1.356. System with common pawl and

a. Lever on button.
b. Slide rod on button.
1 = bevel.   2 = pawl.
3 = lever.

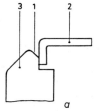

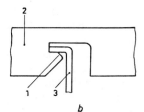

The position of the member driven by the slide rod or lever at its two extreme settings is governed directly by the rod or lever stop in the disengaged state, and by the pawl position with which the rod or lever engages when the button is pressed. Exact register between pawl, pivot and stop position is necessary to ensure a stroke of accurate length.

1.6.5 *Stop mechanisms*

The object of these mechanisms is to detain (or arrest) the sliding member or rotating shaft in a number of preferred positions. This is done by means of a spring-loaded or rotating engaging element, or catch, which drops into a matching slot in the member to be detained. On the approach to a preferred position, the catch helps the moving member to assume this position and, having done so, it then prevents further movement in either direction. The construction is usually symmetrical.

The principle is illustrated diagrammatically in Fig. 1.357, where the forces acting on the two members are indicated without regard to the friction set up in the respective guides. Note the force R acting on the sliding member, and the force F exerted on the catch.

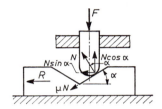

Fig. 1.357. Principle of a stop mechanism.

Ignoring the friction on the slope α, $R = F \tan \alpha$. Because of friction, there is also a force μN, giving rise to a force $\mu N \cos \alpha = \mu F$. Both forces have to be overcome, so

$$R = F(\tan \alpha + \mu) \text{ when } R \text{ drives};$$
$$R = F(\tan \alpha - \mu) \text{ when } F \text{ drives}.$$

Another force R_0 stems from the driven object, making $R_{\text{total}} = F(\tan \alpha + \mu) + R_0$. Once the latch has left the groove, only the forces μF and R_0 remain active, so $R'_{\text{total}} = \mu F + R_0$. Hence we may take

$$\frac{R_{\text{total}}}{R'_{\text{total}}} = 1 + n, \quad \text{where } n = \frac{F \tan \alpha}{\mu F + R_0}$$

With $n > 1$, the catch locates the stop position of its own accord: with $n < 1$, it may have to be helped into this position, perhaps manually. At the same time, n must not be too small, or the mechanism will lack palpable response.

Practical values are

$n = 1{\cdot}2 \ldots 1{\cdot}5$ and $\alpha = 45° \ldots 50°$ for automatic
                                engagement (self-locating).

$n = 0{\cdot}3 \ldots 0{\cdot}5$ and $\alpha = 30° \ldots 40°$ for manual
                                engagement (not self-locating).

Very much the same applies to a rotating shaft. Here, n is the ratio of arresting torque to driving torque of the shaft. A self-locating stop mechanism without any (intermediate) dwell is sometimes stipulated, whereby the stop position cannot be sought independently. This problem has yet to be solved satisfactorily (Fig. 1.358).

Friction can be reduced through the use of a roller, which virtually eliminates μF from the calculations. However, the diameter of the roller must not be too large or it will rest on top of the teeth. Given the slope (α) of the tooth flank and the elevation (h) of the roller on leaving the groove in Fig. 1.359, the maximum roller radius (r) and the minimum roller elevation (h_0) permissible can be calculated. The two values, expressed in terms of the tooth pitch (t), are as follows:

with $\alpha = 30°$: $h_0 = 0{\cdot}134t$ and $r_{max} = t$;

with $\alpha = 45°$: $h_0 = 0.207t$ and $r_{max} = 0{\cdot}707t$.

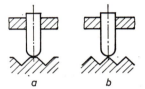

Fig. 1.358. Intermediate positions of a stop mechanism.

a. With intermediate positions.

b. Chance of an intermediate position has been reduced to a minimum.

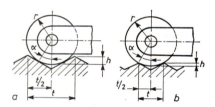

Fig. 1.359. Situation in stop mechanism operating with a roller.

    a. Palpable response.

    b. Roller diameter too large.

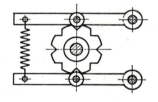

Fig. 1.360. Double stop mechanism in which the two arresting forces cancel one another out.

The shaft can be made to run more lightly by relieving the load imposed by the arresting forces. In Fig. 1.360, the two arresting forces involved are equal. Simple load-relieved stop mechanisms are also shown in Fig. 1.361. Although it does not always roll in the structure, a ball can nevertheless be used very effectively by virtue of its hard, smooth surface (Fig. 1.362). Systems involving a sliding ball and a rotating roller are compared in Fig. 1.363. Practical examples of levers with roller and activated by a spring are given in Fig. 1.364.

Finally, Fig. 1.365 illustrates the principle of a magnetic stop mechanism, shown in the engaged position.

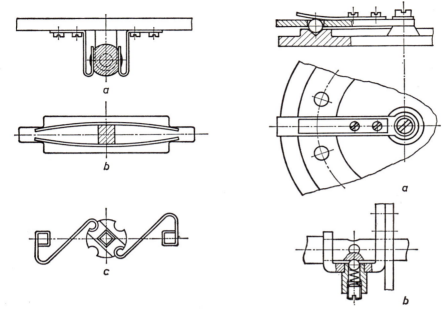

Fig. 1.361.   Fig. 1.362.

Fig. 1.361. Simple load-relief stop mechanism.

a. Flat part of shaft between two flat springs.
b. Square part of shaft between two flat springs.
c. Disc with four slots between two flat springs.

Fig. 1.362. Stop mechanism with ball.

a. Ball under flat spring arrests on ring.
b. Ball under helical spring arrests on shaft.

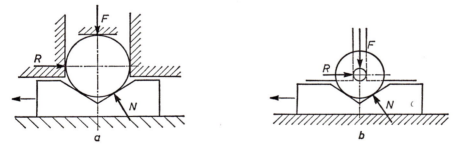

Fig. 1.363. Stop mechanism with ball or roller.

a. The ball does not rotate either slide, because the journal friction is exerted to the circumference of the ball.

b. The roller does roll, because the journal friction is exerted at a small arm.

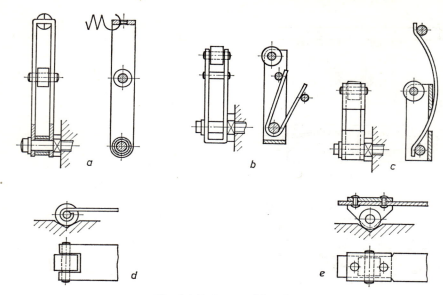

Fig. 1.364. Lever with roller.

a. With tension spring.   c. With flat spring.
b. With hairpin spring.   d and e. Lever acting as spring.

Fig. 1.365. Magnetic stop mechanism
in engaged position.
1 = magnets.

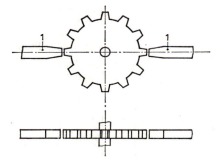

1.6.6 Friction grips

Their task is to retard the motion of structural members by applying one
or more frictional forces to them. Examples are: lid on box (Fig. 1.366);
sliding shaft with friction grip (Fig. 1.367); sliding bracket on rectangular
strip (Fig. 1.368). The elastic element of a plug-and-socket construction is
difficult to accommodate in the plug without causing it to be permanently
deformed when in use (see Fig. 1.369). It is easier to make the socket resilient.
Plug pins with slots sawn in them are unsuitable.

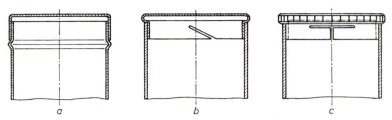

Fig. 1.366. Lid held on box by friction.

a. Friction set up by slight out-of-roundness.
b. Made springy by saw-cut.
c. Friction set up by spring tabs.

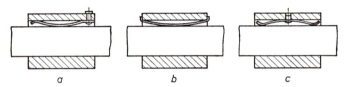

Fig. 1.367. Sliding shaft with friction grip.

a. Flat spring riveted on.
b. Flat spring interposed with ends bent up.
c. Flat spring held by riveted plug.

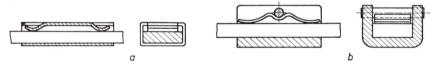

Fig. 1.368. Sliding bracket on strip.

a. Bracket has two spring tabs.   b. With extra flat spring.

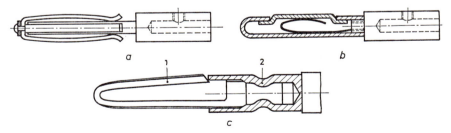

Fig. 1.369. Spring plug pin.

a. With two flat springs for four contact positions riveted on (banana plug).
b. Plug with separate member held in gap by pressure of hairpin spring.
c. Hook-shaped plug pin of punched and hardened beryllium copper.
    1 = plug pin.   2 = copper holder.

1.6.7 Stops[7, 8]

These devices restrict the movements of rotating or sliding members. A stop may be fixed, as in Figs. 1.370 and 1.371, or adjustable, as in Fig. 1.372. With an adjustable stop the length of the stroke can be varied as required, as, for example, in the case of office machines.

Fig. 1.373 illustrates a method of limiting a rotating roll of paper by means of easily adjustable stops.

The collision occurring when the moving member reaches the stop may cause a certain amount of shock, depending on the speed of travel. Particularly

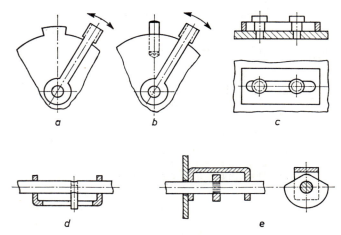

Fig. 1.370. Simple fixed stop.

a. Fixed lug on disc limits movement of rotating lever.
b. Fixed pin in disc limits movement of rotating lever.
c. Two shouldered screws limit sliding movement.
d. Pin in slot limits stroke of sliding shaft.
e. Cam on shaft, together with channel bracket, limits both
    sliding and rotary motion of shaft.

in the case of rapid action mechanisms, must the reverberation of members brought to a sudden stop be taken into consideration. The shock can be lessened by employing an elastic stop (Fig. 1.374). Beyond this stop, the movement is finally arrested by another, fixed stop. However, moving, elastic stops are bulky and often excessively inert, so a fixed stop is preferred. This should be resistant to wear, and made of a shock-absorbing material: the possibility that its properties will change with varying temperature should also be taken into account. The plastics, polyurethane rubber (Vulkollan), polyamide (nylon) and polyacetal (Delrin) are suitable for the purpose.

Sometimes, a movable stop must be adjustable, as must the stops that govern the winding width of coil-winding machines (Fig. 1.375).

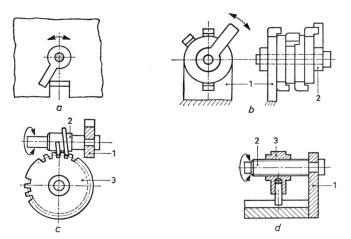

Fig. 1.371. Fixed stop for wide angle of rotation.

a. Maximum angle of rotation is slightly less than 360°.
b. Shaft with lug (2) can perform a number of full revolutions relative to frame (1); each intermediate disc permits rotation through an angle, as in a, until the flanged lug is checked by the lug on the adjacent disc.
c. Angle of rotation of worm shaft (2) in relation to frame (1) is limited by worm wheel (3) from which several tooth spaces have been omitted.
d. Angle of rotation of lead screw (2) relative to frame (1) is governed by the limited stroke of nut (3).

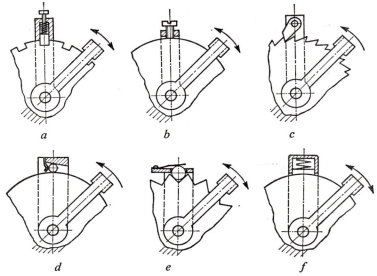

Fig. 1.372. Stop adjustable by means of:

a. Catch.   c. Ratchet and pawl.   e. Stop mechanism.
b. Clamping screw.   d. Friction ratchet.   f. Brake.

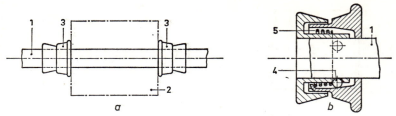

Fig. 1.373. Stops to retain roll of paper on shaft.

a. Shaft with stops.   *b.* Enlarged cross-section of stop.
1 = shaft.   2 = roll.   3 = stop.
4 = three balls distributed round the circumference.   5 = helical spring.

The helical spring presses the separate halves of the stop together, thereby forcing the balls to ride up the conical sides of the recess and press tightly against the shaft, thus locking the stop onto the shaft. The stop is released by pushing the separate halves of the stop apart.

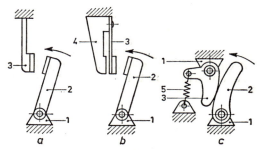

Fig. 1.374. Elastic stop.

a. Without fixed stop.
b. With following fixed stop.
c. Elastic stop, which gradually engages a fixed stop.
1 = frame.   2 = moving part.   3 = elastic stop.
4 = fixed stop.   5 = spring.

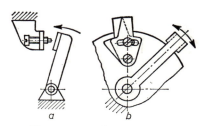

Fig. 1.375. Adjustable stop.

a. With adjusting screw.
b. With adjustable finger, which can be fixed by means of screw.

REFERENCES

[1] RICHTER, V. VOSS and KOZER, *Bauelemente der Feinmechanik*, Verlag Technik, Berlin, 1954.

[2] K. HAIN, *Die Feinwerktechnik*, Fachbuch Verlag Dr. Pfanneberg, Giessen, 1953.

[3] G. SCHLEE, *Feinmechanische Bauteile*, Verlag Konrad Wittwer, Stuttgart, 1950.

[4] D. C. GREENWOOD, *Product Engineering Design Manual*, McGraw-Hill, New York/Toronto/London, 1959.

[5] K. RABE, *Grundlagen Feinmechanischer Konstruktionen*, Anton Ziemsen Verlag, Wittenberg/Lutherstadt, 1942.

[6] N. P. CHIRONIS, *Machine Devices and Instrumentation*, McGraw-Hill, New York/London, 1966.

[7] K. RABE, *Konstruktion und Berechnung mittelbarer Festanschläge*, Feinwerktechnik, 1961, 8, p. 277.

[8] K. RABE, *Anschläge, eine Untergruppe der Sperrungen*, Feinwerktechnik, 1961, 5, p. 166.

[9] S. B. TUTTLE, *Latch and Trip Mechanisms*, Machine Design, 1967, December 7, p. 179.

[10] G. MUTZ, *Dämpfung von Prellungen durch einseitige Impulstranslation*, Feinwerktechnik, 1966, 12, p. 581.

[11] S. HILDEBRAND, *Feinmechanische Bauelemente*, 1968, Carl Hanser Verlag, München.

1.7 Springs[1−5, 9, 27, 28, 31]

1.7.1 Introduction

For particulars of spring materials, see Volume 2, Section 1.5. For design theory, see Volume 1, Section 2.2 and for particulars of electrical contact springs Volume 1, Chapter 5.

By virtue of elastic deformation, springs can absorb or supply energy and thus produce tensile or compressive forces, bending moments or torque. Springs are used as:

- Energy accumulators, e.g. for clocks and toys.
- A means of restoring to their stable state members that have in some way been deflected from it, as in valves, weighing machines and measuring instruments.
- Buffers in vehicle shock absorbers.
- Elastic joints or fasteners.
- Hinges in backlash-free bearings, where the angular rotation required is small.

In designing a spring, the aim is to satisfy the requirements imposed, whilst using only a small amount of material. This can be done, if the spring absorbs a great deal of energy per unit of volume. Given the stress and the modulus of elasticity of the spring, the form factor is a quantity depending on the shape of the spring and the kind of load imposed on it (see Volume 1, Section 2.2.7).

The factor increases with the amount of material placed under maximum stress, and is highest in a taut wire, i.e. 0·5. In torsion and flexural springs, only the outer fibre is subjected to maximum stress, and the factor is less than 0·5. The result obtained with such springs depends also on the shape of the cross-section under load.

With flexural springs, where the moment varies from one part of the spring to another, the moment of inertia must also vary, to produce a high factor.

The material primarily employed is hardened spring steel strip or wire, where this can be wound or otherwise shaped to produce the spring required. Flat springs, which have to undergo substantial deformation in manufacture, are shaped in unhardened steel, and afterwards hardened and (if necessary) tempered.

When a non-magnetic material is required, or if the spring is also to serve as an electrical conductor, phosphor bronze, German silver, tombac, brass and other materials of this kind are employed. They have acquired the necessary stiffness as a result of being drawn or rolled. In certain cases, special materials are employed, such as the nickel steels, Invar and Elinvar, or beryllium copper, hard or of hardening quality.

Some flat springs, for example the main springs of clocks, which are wound into spiral form, are made of closely toleranced round wire, rolled flat before the actual spring is formed. This saves the labour of cutting short lengths of strip, and also results in rounded edges that reduce friction in the spring barrel.

Rubber or plastics is sometimes used as spring material, but only when there is no specific spring characteristic to be complied with. Special attention must also be given to the environment of such springs: temperature, light, air (oxygen and ozone) and contact with copper or copper alloys.

The permissible stress in the spring material depends, not only on the nature of the material, but also on:

● The nature of the load (static, shock or alternating).
● The number of load changes involved in shock or alternate loading.
● The stresses already present in raw material not stress-relieved.
● The surface condition of the material.
● The shape of the spring, which may involve stresses set up in the material by winding, bending or punching.
● Abrupt changes of cross-sectional area, which may have a notch effect in hard material.
● Probable corrosion.

With these factors in view, the acceptable stress for flexed springs is limited to not more than 0·7 the maximum permissible for the unprocessed raw material.

1.7.2 Flexural springs[6, 7, 30]

For a spring of a given length and material, the choice of cross-section will be governed by the load and the flexure. If both are substantial, a rectangular cross-section is required, very much wider than it is deep: as for a clockwork spring, in view of its great length, or for a vehicle spring made up of several narrow springs stacked one on top of the other. Where the load and flexure involved are small, a round cross-section will do, as, for instance, in the case of a spring pointer stop.

From the point of view of shaping, very much more can be done with flat material than with wire, which can only be subjected to progressive bending. The best springs are obtained with material that does not have to be bent in manufacture or, better still, need not be shaped at all. The rolling direction of flat springs is immaterial, provided no sharp bends, or rounded bends with only a small radius, are made in them. Where such bends do occur, they are taken at right-angles to the direction of the rolling operation, and the minimum permissible radius of curvature adhered to, at any rate where the (maximum) bending stress occurs. When two bends perpendicular to each other are required, they should be at 45° to the rolling direction.

(a) *Straight flat springs* are used for small, almost rectilinear deflections, and for a given variation of force within a small stroke. In view of the straight shape, it is best to use spring material with a yield point just below its tensile strength: the manufacture of such a spring is relatively simple, since it does not involve a bending operation. Flat springs mounted under initial stress are shown in Fig. 1.376. The bending stress is greatest where clamping begins, so it is better to give a slight radius to the edge of the cover plate, over which the spring bends, to counteract undesirable stress concentrations. Undue weakening of the spring, as by fixing holes, should be avoided.

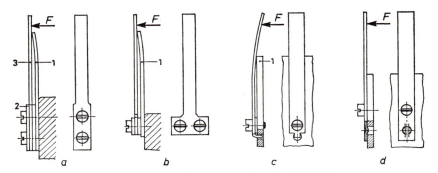

Fig. 1.376. Flat springs mounted under initial stress.
a. Spring widens near the fixing holes.
b. Undue shape of spring material.
c. Spring attached by one screw, with tab to prevent twisting.
d. Spring attached by one screw, with dimple to prevent twisting.
1 = supporting plate.   2 = cover plate.   3 = flat spring.

The initial stress of the spring is supplied by a supporting or backing plate, which also gives scope for adjustment of the spring working point, an advantage in the case of stacked switches with small contact gaps and stroke.

Stacks of this kind, with the contact springs isolated from one another, are shown in Fig. 1.377. For a given length of spring, the spring characteristic is governed by the (varying) width.

To limit the cost of the blanking tool and the material consumption, flat springs, clamped parallel to one another, are usually made equal in width over their full length. The cross-sectional area of the spring is reduced as much as possible towards the free end in order to obtain:

- More uniform loading of the spring material and thus avoid any pronounced, critical cross-section.
- A more flexible spring (flatter spring characteristic) at a given acceptable bending stress.
- A smaller mass action in the case of fast-moving springs; the mass is then smaller at the precise point where the amplitude is greatest.
- Reduced sensitivity to disturbances from outside.

To accommodate a contact despite this, the spring must have a sufficiently wide end. With the object of matching the spring characteristic as closely as possible to that of a triangular spring, whose elastic line is circular, tapered sides (or a triangular hole with rounded corners) are used to avoid undue stress concentrations (Fig. 1.378). With parallel contacts on contact springs, the contacts should be made to come together independently, as far as possible: this makes good contact between them more likely. Three different versions are shown in Fig. 1.379 (see also Volume 1, Section 2.2.6).

The springs can be actuated by a controlling member made of insulating material. Figure 1.380 illustrates one or two practical combinations of spring and controlling member.

Parallel motion accompanied by only very slight lateral displacement can be obtained with the system of Fig. 1.381, in which two parallel flat springs are employed.

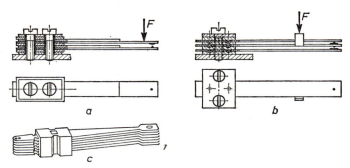

Fig. 1.377. Spring assembly of stacked switch with isolated contact springs:

a. With two screws passed through insulating bushes.
b. With two screws alongside the spring assembly: the springs are dimpled slightly to prevent twisting.
c. By embedding in thermoplastic.

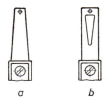

Fig. 1.378. Nearest approach to flat spring of triangular shape.

a. Tapered on both sides.

b. Provided with triangular punched hole with rounded corners.

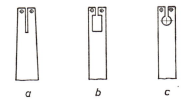

Fig. 1.379. Flat spring for two parallel contacts.

a. Spring with split end.

b. Control member can be passed through square hole.

c. Round hole is preferable as a means of avoiding stress concentrations.

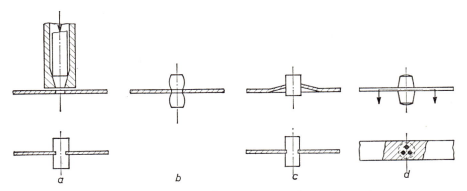

Fig. 1.380. Attachment of actuating member to flat spring.

a. Member of hard rubber, heated and then pressed into hole hollowed out in spring.

b. Rubber actuating member placed in hole and then expanded at a raised temperature.

c. Member subjected to plastic deformation together with spring.

d. Member moulded on and held in place by three holes.

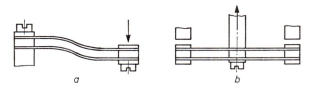

Fig. 1.381. Parallel motion with only slight lateral displacement

a. One contact on two flat springs.

b. Two contacts moved by pin via two flat springs.

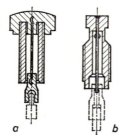
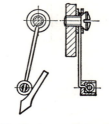

Fig. 1.382. Spring system of top bearing in electricity supply meter.
a. Wire spring embedded at both ends.
b. Wire spring soldered in acts also as journal for bearing.

Fig. 1.383. Spring pointer-stop.

(b) *Straight wire springs* are used where there is need for some elasticity in all the cross directions and only small forces occur. Figure 1.382 gives two versions of a spring collar bearing for the rotor of an electricity supply meter. Figure 1.383 shows a spring pointer-stop.

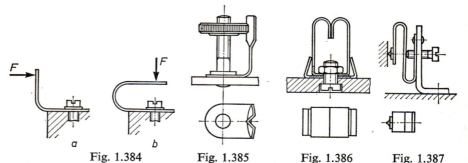

Fig. 1.384   Fig. 1.385   Fig. 1.386   Fig. 1.387

Fig. 1.384. Principal forms of bent flat springs.
        a. Spring bent at right-angles.   *b.* Spring bent into U-shape.
Fig. 1.385. Bent flat spring used to lock knurled nut.
Fig. 1.386. Bent flat spring used as contact spring. The U-shaped auxiliary spring increases the contact force.
Fig. 1.387. Flat spring bent into S-shape can be adjusted to a given contact force by the set-screw.

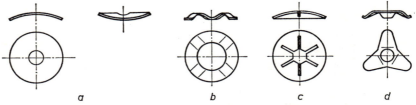

Fig. 1.388. Four types of flexural spring with steep characteristics.
    a. Saddle spring.   *c.* Bell-mouthed (shaped) spring.
    b. Crinkle washer.   *d.* Star-shaped spring.

(c) *Bent flexural springs* are employed instead of straight springs when there is not enough room for a straight spring; an existing point of attachment in another plane is to be used; attachment in another plane is necessary for assembly reasons.

Fig. 1.384 shows two of the principal forms of these springs, whilst applications are depicted in Figs. 1.385 to 1.387 inclusive. The springs shown in Fig. 1.388 are used for locking, for axial retention without backlash, and to supply a certain backing pressure: they are essentially flexural springs, and have a steep spring characteristic, so cannot provide precisely-defined forces. Fig. 1.389 shows a compression spring accommodated in a small space for a push-button switch.

The bellows of Fig. 1.390 is encountered in pneumatic devices in various materials from commercial sources: ether capsules consisting of two corrugated walls (Fig. 1.391) are employed in barometers. Fig. 1.392 shows an elastic tube used to measure pressure in gas or liquid. The shape of these springs gives them a minimum spring constant in the specific direction in which they move in operation.

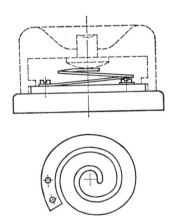

Fig. 1.389. Spiral compression spring in push-button.

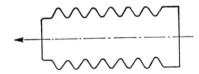

Fig. 1.390. Bellows for pneumatic devices.

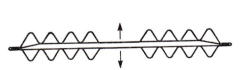

Fig. 1.391. Ether capsule for barometer.

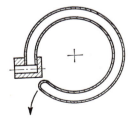

Fig. 1.392. Elastic tube for manometer.

(d) *Coiled strip springs*[13, 14] are wound flexural springs made of rectangular cross-sectional material. They are used: as driving springs, e.g. in clocks and toys; as balance springs in clocks; and as a means of supplying restoring torque in measuring instruments.

In cheap clocks, the outer end of the driving spring (or main spring) is usually anchored at a fixed point, and the spring wound up from inside: a ratchet wheel on the shaft isolates the winding action from the remainder of the clock movement. In more expensive clocks, the main spring is enclosed in a barrel: its outer end is fastened to the rim of the barrel, and its inner end coupled direct to the movement via the shaft. Winding takes place from outside, with a ratchet wheel between barrel and plate. The inner end of such a spring can be anchored in the manner of Fig. 1.393. Where a barrel is provided, the outer end is anchored as in Fig. 1.394; otherwise according to Fig. 1.395 with, say, an eyelet round a pin.

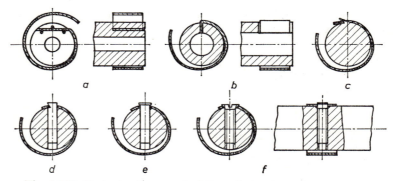

Fig. 1.393. Fastening inner end of flat spiral strip spring to arbor.

a. In axial slot, afterwards closed.   *d.* On pin pressed into arbor.
b. In radial slot.   *e.* On pin, with head, pressed into arbor.
c. On projection.   *f.* On constricted head of countersunk screw.

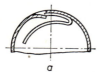 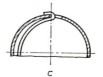

Fig. 1.394. Fastening outer end of flat spiral strip spring to spring barrel.
a. End of spring bent-up and hooked behind tab.
b. Hole in end of spring hooks on to tab.
c. End of spring bent up and hooked into hole.

Fig. 1.395. Fastening outer end of flat spiral strip spring without barrel.

The fastening of the outer ends of balance springs (or hairsprings, as they are also called) must be easy to adjust within certain limits, to allow for regulation of the clock movement. This is accomplished by a tapered pin (Fig. 1.396) or a clamping screw (Fig. 1.397).

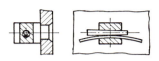

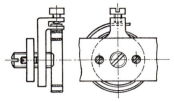

Fig. 1.396. Outer end of balance spring fastened by a tapered pin.

Fig. 1.397. Outer end of balance spring fastened with clamping screw in adjustable bracket.

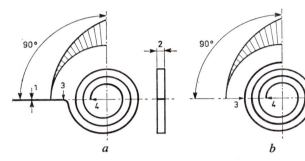

a   *b*

Fig. 1.398. Flat spiral spring to supply restoring torque in measuring instruments.
    a. Spring winds itself up under load.   *b.* Spring unwinds under load.
Thickness (1) is about one tenth of width (2) of strip. Spring is measured between the two points (3) and (4).

The second method eliminates the tendency of the spring to slip round during assembly, and makes regulation simpler than in the cheaper clock with its tapered pin.

Flat spiral springs, to supply restoring torque in measuring instruments, are made of bronze and are obtainable for torque values from 0.025 to 110 μN m with 90° angular displacement (Fig. 1.398).

The ends of the spring are usually anchored by soldering, particularly where the spring serves also as an electrical conductor.

Clamping or screw-fastening is employed mostly for slightly larger springs or when the spring length has to be adjustable, say, to facilitate calibration of the instrument concerned. A soldered joint also gives scope for adjustment, if the end of the spring is gripped in an elastic clamp before soldering.

(e) *Coiled wire springs*[26] are coiled flexural springs made of round material. They are used a lot in instrument manufacture to restore moving parts to the rest position after they have been deflected through an angle. They are also known as torsion springs.

To minimize the load on the arbor, it is best to position the free ends of the spring as far as possible parallel to one another and on the same side of the arbor under the maximum stress (Fig. 1.399*a*). A very unfavourable example is illustrated in Fig. 1.399*b* where the resultant of both spring forces F_1 and F_2 almost amounts to their sum.

With powerful springs in a large spring space, double arbor bearings or seats are recommended, to prevent excessive load deflection (Fig. 1.400). In Fig. 1.401, one end of the spring is inserted in the crank and the other end hooked round a pin in the arbor.

Highly specialized applications of coiled wire torsion springs are illustrated in Fig. 1.402, showing a spring-loaded override snap coupling, and Fig. 1.403, for a clamping spring. In the latter mechanism, the inner diameter of the spring is so dimensioned, at about 0·3 mm smaller than the shaft diameter, that the shaft encounters very little frictional resistance when rotating in the direction indicated by the arrow, but generates more and more friction owing to the constricting coils of the spring when rotating in the opposite direction. In other words, the spring acts as a brake.

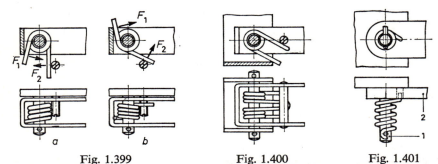

Fig. 1.399   Fig. 1.400   Fig. 1.401

Fig. 1.399. Method of fitting a torsion spring, taking bearing forces into account.
    a. Preferred structure.
    b. Non-preferred structure.
Fig. 1.400. Double bearing for powerful spring in large spring space.
Fig. 1.401. Torsion spring engaging arbor (1) and lever (2).

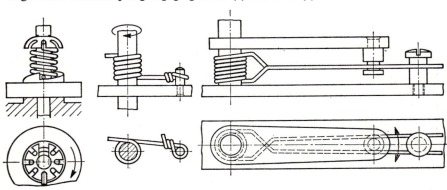

Fig. 1.402   Fig. 1.403   Fig. 1.404

Fig. 1.402. Spring-loaded override snap coupling: disc can rotate only in direction indicated by arrow.
Fig. 1.403. Clamping torsion spring on round shaft; shaft can rotate only in direction indicated by arrow.
Fig. 1.404. Torsion spring with two-way action.

The two-way action of the wire torsion spring in Fig. 1.404 gives the crank a very clearly defined centre position, whence it can pivot through a given angle in either direction, against the increasing pressure of one of the two free ends of the spring.

1.7.3 Torsional (shear stress) springs[8, 15-17, 29]

Straight torsional springs are not very much used in precision engineering, because they take up a fair amount of room and are difficult to anchor. This type of spring is sometimes encountered as an elastic hinge pin in lightweight cover structures; as the spring element in a torsion bar bearing, as shown earlier in Fig. 1.325; or in the elastic coupling of Fig. 1.405.

Helical *coiled torsional springs* are much used as compression or tension springs. Such helical springs are usually made of round material (spring steel wire), but sometimes also of material of rectangular cross-section (spring steel strip). The advantages of springs of this kind are that they are very flexible and are able to carry heavy loads in a small space. For particulars of calculations relating to them, see Volume 1, Section 2.2.6.

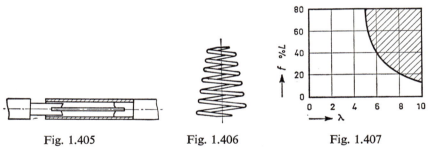

| Fig. 1.405 | Fig. 1.406 | Fig. 1.407 |

Fig. 1.405. Elastic coupling with flat spring as torsion bar.
Fig. 1.406. Conical (or tapered) torsional spring.
Fig. 1.407. Diagram for determining the buckling of helical coiled torsional springs, for compression springs with flat end faces, supported on parallel faces perpendicular to the spring axis. λ = slenderness of the spring, that is, the ratio of length of spring in its relaxed state to the diameter of the spring: f is expressed as a percentage of the relaxed spring length (L). In the hatched zone, the spring buckles.

Most *compression springs* are cylindrical. Where greater stiffness against transverse buckling is required, the conical coiled spring (Fig. 1.406) can be employed. A cylindrical spring is prevented from buckling by seating it in a cylindrical recess, or round a pin, although both methods involve frictional losses.

It is better to avoid making the length-to-diameter ratio (slenderness) too large when dimensioning the spring (Fig. 1.407), and to match the ends of the spring as exactly as possible to the parallel bearing faces at right-angles to the spring axis (Fig. 1.408). A plain end of clipped wire is cheap but unfavourable, and really acceptable only in the case of thin wire. Squared ends cost slightly more; the bearing face is favourable and the wires cannot become entangled in transit. Ground ends, squared or plain, are fairly expensive and difficult to produce on thin wire springs. The area of contact may cover nearly 360°.

In the case of *tension springs*, the tensile forces should be concentrated as much as possible at the centre of the spring. To ensure this, the spring ends are looped to provide suitable anchorage points. In forming the eyes, or loops, consideration must be given to the smallest permissible bending radius, since these parts usually govern the strength of the spring. Examples of formed spring loops are given in Fig. 1.409. The loop can simply be hooked into a hole cut in another part of the structure, or fastened in one of the ways illustrated in Fig. 1.410. Tension springs without loops can be fastened as in Fig. 1.411.

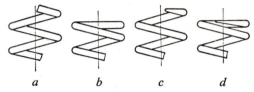

Fig. 1.408. Ways of finishing the ends of helical coiled torsional springs (compression springs).

a. Plain clipped ends: all the turns are active.
b. Ends squared: number of active turns is two fewer than the total number.
c. Ends ground: number of active turns is one fewer than the total number.
d. Ends ground and squared: number of active turns is three fewer than the total number.

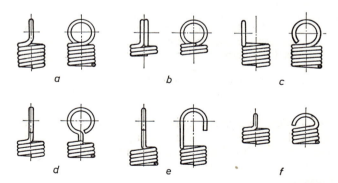

Fig. 1.409. Loops formed on helical coiled torsional springs (tension springs).

a. Single full loop over centre.	*d.* Extended full loop over centre.
b. Double full loop over centre.	*e.* Long, round-end hook over centre .
c. Full loop at side.	*f.* Half loop.

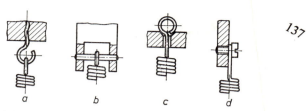

Fig. 1.410. Methods of fastening looped tension springs

a. Embedded hook.   *c.* Loop bears on rim of countersunk
b. Pin pressed in.   *d.* Cheese-head screw passed through

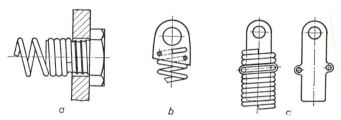

Fig. 1.411. Methods of fastening tension springs without loops.

a. End of spring fits around screw. An ordinary screw can sometimes be used.
b. One-and-a-half turns of spring in three holes in plate.
c. One of the turns passes through two holes in the strip: the active length of the spring can be adjusted.

1.7.4 Springs with non-linear characteristics[10 − 12, 18, 19, 25]

Some types of spring, when suitably installed, have a characteristic that is non-linear: that is, it consists of two or more straight sections. Examples of such non-linear helical springs are given in Fig. 1.412, and of non-linear flat springs in Fig. 1.413. With these springs, there is an extra-linear increase in force with deflection. Springs having the opposite effect, and therefore a flatter characteristic, are shown in Fig. 1.414: when they are employed so as to utilize only a fraction of their full deflection, an approximately flat portion of the characteristic can be exploited.

Conical disc springs (Belleville washers), shown in Fig. 1.415, take the form of a truncated cone with an apex angle of about 165° and, for the optimum characteristic, a ratio D_u/D_i of about 2. As will be seen from Fig. 1.416, the associated curves have a maximum. In this zone the characteristics are more or less flat, so this type of spring can be used to supply fairly constant pressure, where the (tolerance governed) space available for the spring element is small, as with friction clutches and elastic retaining and sealing devices. To obtain substantially greater deflection, several of the springs are stacked in a resilient column (nest) capable of absorbing a lot of energy in a small space. Various relationships between the deflection and the resultant moment or force can also be obtained in structures fitted with springs having linear characteristics (Fig. 1.417).

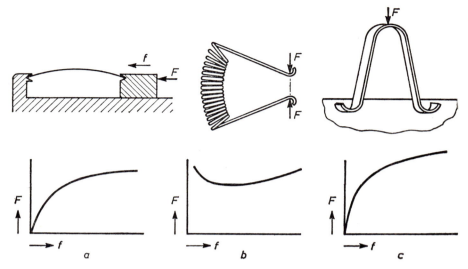

Fig. 1.414. Springs having a sub-linear increase in force with deflection.

a. Bending moment increases with deflection.
b. As helical spring bends, the arm, on which force (F) acts, lengthens at first and afterwards shortens.
c. With deflection, the arm, on which the reaction forces of the fulcrums act, lengthens.

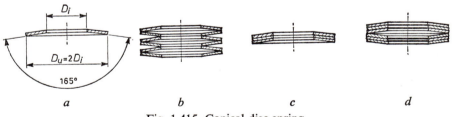

Fig. 1.415. Conical disc spring.

a. Cross-section.
b. The compression by a force is six times that of a single spring.
c. Three times the force is required to produce the same compression as with a single spring.
d. Combination in which the force is three times, and the compression twice, that of a single spring.

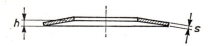

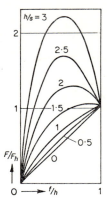

Fig. 1.416. Spring characteristics of differently dimensioned conical disc springs.

h = initial cone height (camber) of spring.
s = thickness of spring material.
With $h/s = \sqrt{2}$, the characteristic is almost horizontal
for $f/h = 1$, that is, in the flattened state.
With $h/s = 0$ (flat ring), the characteristic is straight.

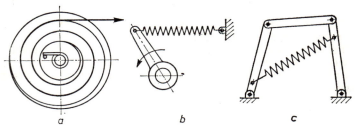

Fig. 1.417. Structure having a spring with a linear characteristic, which nevertheless
produces a non-linear effect.

a. Cable dragging on spiral-shaped drum, in combination with torsion spring,
   results in flat characteristic.
b. Flat characteristic due to the fact that, as the force in the spring increases,
   the arm shortens on which the spring acts.
c. Flat characteristic due to displacement of both ends of spring.

1.7.5 Hinge springs[20–23]

Notable examples of the use of (flat) springs as hinges will be found as
follows:

Flexural hinge springs in Volume 7, Section 2.3.2, Fig. 2.11, and also in
   this volume, Fig. 1.279, Fig. 1.418a, b and c and Fig. 1.594c.
Crossed hinge springs in Fig. 1.418d, e and f of this volume, and in Volume 7,
   Section 2.3.2, Fig. 2.12.
Torsion hinge springs in Section 1.5.6(c), Fig. 1.325 herein.
Parallel-motion in Section 1.7.2(a), Fig. 1.381 herein.
The conversion of sliding motion to rotary motion, Fig. 1.419 herein.
Hinge motion in the transmission system of a probe, Fig. 1.420 herein.

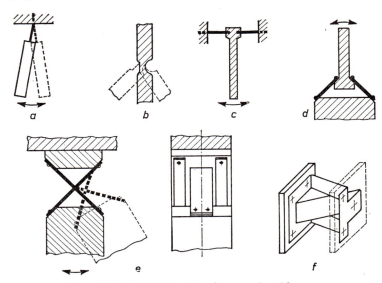

Fig. 1.418. Principles of springs used as hinges.

a. Flat spring as flexural hinge spring.
b. One-piece polypropylene hinge.
c. Two flat springs forming a flexural spring hinge.

d. Two flat springs forming crossed spring hinge.
e. Three flat springs forming crossed spring hinge.
f. One-piece crossed spring hinge.

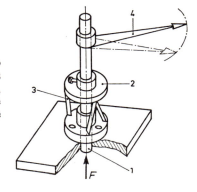

Fig. 1.419. Conversion of sliding motion to rotary motion. As pin (1) rises, disc (2) rotates through the action of the three flat springs (3), and pointer (4) indicates the stroke of the pin. The three flat springs also supply the restoring torque.

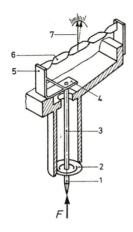

Fig. 1.420. Transmission system of a probe. As the probe (1) rises, guided by diaphragm (2), rod (3) forces spring (4) up, thereby dragging spring (5) to the left. This puts a tension load on torsion spring (6) and makes it rotate, taking pointer (7) with it, to indicate the stroke or displacement of the probe.

REFERENCES

[1] RICHTER, V. VOSS and KOZER, *Bauelemente der Feinmechanik*, Verlag Technik, Berlin, 1954.

[2] K. HAIN, *Die Feinwerktechnik*, Fachbuch Verlag Dr. Pfanneberg, Giessen, 1953.

[3] G. SCHLEE, *Feinmechanische Bauteile*, Verlag Konrad Wittwer, Stuttgart, 1950.

[4] D. C. GREENWOOD, *Product Engineering Design Manual*, McGraw-Hill, New York/Toronto/London, 1959.

[5] N. P. CHIRONIS, *Machine Devices and Instrumentation*, McGraw-Hill, New York/London, 1966.

[6] *Blattfedern und deren Einsatz im Gerätebau*, Microtechnik, 1966, April, No. 2.

[7] J. PALM and K. THOMAS, *Berechnung gekrümmter Biegefedern*, VDI-Zeitschrift, 1959, No. 8.

[8] K. H. BERG, *Konstruktionsbeispiele aus der Feinwerktechnik*, Konstruktion, 1956, 5, p. 187.

[9] F. WOLF, *Die Federn im feinmechanischen Geräte und Instrumentenbau*, Deutscher Fachzeitschriften- und Fachbuchverlag, Stuttgart, 1953.

[10] K. H. WALZ, *Die Tellerfeder im Machine- und Werkzeugbau*, Blech, 1959, 4, p. 171.

[11] K. H. WALZ, *Entwurf und Konstruktion der Tellerfeder*, Werkstatt und Betrieb, 1957, 5, p. 311.

[12] W. WERNITZ, *Die Tellerfeder*, Konstruktion, 1954, 10, p. 361.

[13] *Spiralfedern für Betriebsmessinstrumente*, DIN 43801.

[14] E. JOERG, *Die Spiralfeder als Antriebmittel in Federwerken und Uhren*, Feinwerktechnik, 1954, 3, p. 81.

[15] W. A. WOLF, *Die Schraubenfedern*, Verlag W. Girardet, Essen, 1950.

[16] *Zylindrische Schraubenfedern aus runden Drähten und Stäben*, DIN 2089; Drehstabfedern, DIN 2091.

[17] S. HILDEBRAND, *Zur Berechnung von Torsionsbänder im Feingerätebau*, Feinwerktechnik, 1957, 6, p. 191.

[18] E. H. BOERNER, *Constant Force Compression Springs*, Product Engineering, 1954, September, p. 129.

[19] F. A. VOTTA, *Constant-Force Springs*, Machine Design, 1963, January 31, p. 102.

[20] H. STABE, *Federgelenke im Messgerätebau*, VDI-Zeitschrift, 1939, No. 45, p. 1189.

[21] W. WUEST, *Blattfedergelenke für Messgeräte*, Feinwerktechnik, 1950, 7, p. 167.

[22] H. ANGERMAIER, *Federgelenke und Federgelenkgetriebe, Bauelemente der Feinwerktechnik*, Feinwerktechnik, 1966, 12, p. 553.

[23] K. H. SIEKER, *Federgelenke und Federnde Getriebe*, VDI-Tagungsheft 1 Getriebetechnik, Düsseldorf, VDI-Verlag, 1953, p. 141.

[24] D. C. GREENWOOD, *Engineering Data for Product Design*, McGraw-Hill, New York/Toronto/London, 1961.

[25] E. F. GÖBEL, *Gummi und seine Anwendung in der Feinwerktechnik*, Feinwerktechnik, 1962, 8, p. 291.

[26] *Gewundene Biegefedern*, DIN 2088.

[27] N. P. CHIRONIS, *Spring Design and Application*, McGraw-Hill, New York/London, 1961.

[28] A. M. WAHL, *Mechanical Springs*, McGraw-Hill, New York/London, 1963.

[29] J. H. KEYES, *Nomographic Design of Helical Springs*, Machine Design, 1950, December, p. 173.

[30] G. VÖLKERLING, *Zur Berechnung von Kontaktfedern der Feinwerktechnik*, Technica, 1966, 16, p. 1479.

[31] S. HILDEBRAND, *Feinmechanische Bauelemente*, 1968, Carl Hanser Verlag, Munich.

1.8 Controls[1−3]

Most controls, particularly those of precision engineering mechanisms, are operated by hand: pedal controls are rare, and will be ignored. Hand force can be exerted on: push or pull grips or handles; slide grips, mainly for linear motion; rotary grips or knobs; cranks; keys; carrying handles.

The controls are usually connected permanently to the driven member, although some rotary grips, cranks and keys are not, but are clipped on and off as required.

The shape, size and method of fastening are suited to the requirements imposed: to fit comfortably in the hand, respond readily and be suitable for prolonged operation without causing fatigue. There must be no sharp corners, edges or knurls, and all surface junctions must be given the optimum radius. The dimensions are governed mainly by the forces or moments to be transmitted, and the precision of movement required of the controlling member (ergonometry).

If the control serves also as a regulator, it can be fitted with a position indicator in one of two main forms: with a stationary scale; or with a movable scale (Fig. 1.421).

● Points to consider when choosing the material are:
● Frequent handling soon results in local wear.
● Skin secretion may cause oxidation of metal controls.
● Because they are good thermal conductors, large metal handles, and so on, feel unpleasantly cold.
● Metal controls are plated with, say, chromium, or anodized if made of aluminium. Plastics are often used.

Pull controls are rarely used: circular or semicircular forms occur as rings and as pull-knobs on car dashboards.

Push controls usually take the form of push buttons for rectilinear motion, or keys for rotary or tilting motion, where the forces exerted are small. Where greater forces are exerted, the handle version is generally chosen instead. For fingertip operation, the top of the button or key (often cylindrical in small and cheap devices) is hollowed out or ridged slightly to prevent the finger slipping (Fig. 1.422). Keys intended for rapid movement, as with telegraph signalling instruments, have to be kept under perfect control by the thumb and two or three fingers (Fig. 1.423). To fit comfortably in the palm of the hand, such a key for a wrap-around grip is mushroom shaped (Fig. 1.424).

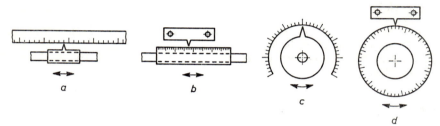

Fig. 1.421. Stationary and movable scales or dials.

a. Straight fixed scale.   c. Round fixed dial.
b. Straight movable scale.   d. Round movable dial.

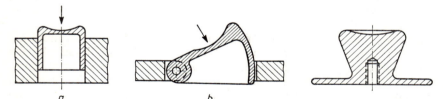

Fig. 1.422. Push controls. a. Push button. b. Key.   Fig. 1.423 Push button or key for rapid movement.

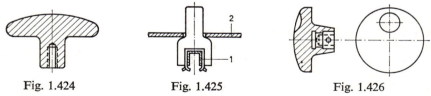

Fig. 1.424   Fig. 1.425   Fig. 1.426

Fig. 1.424. Key for wrap-around grasp.
Fig. 1.425. Slide knob. 1 = guide rod. 2 = cover plate.
Fig. 1.426. Knob with cup-shaped depression for fingertip operation (contact grasp).

Slide grips are used to operate slide potentiometers, or, in a smaller form, as slide knobs for switches, etc. The actuating force should preferably be exerted as close to the guide as possible, as a precaution against sticking (see Section 1.4.2). In Fig. 1.425, the knob slides at right-angles to the plane of the drawing.

Rotary grips are the most used in precision mechanism. They are more-or-less encompassed by the fingers, or grasped by the whole hand, during operation and are employed for rotation through small angles or for several full turns. A large diameter makes it possible to apply a considerable torque, and also obtain accurate adjustment.

The finger knob of Fig. 1.426 has a cup-shaped depression into which a fingertip can be inserted to rotate the knob. Examples of finger knobs with knurled rims are given in Fig. 1.427. The knobs in Fig. 1.428 are intended for small instruments or devices, and are suitable also for push–pull operation.

The rotating knob of unusual shape, shown in Fig. 1.429, is used for electric switches, where the torque to be transmitted may be quite large. It is not suitable for vernier adjustment.

The knobs are fastened to the control spindle by a screw or a spring clip, as shown in Fig. 1.430, whilst Fig. 1.431 shows an assembly in which the fixing screw can be relaxed only a few turns in order to detach the knob, and is therefore captive.

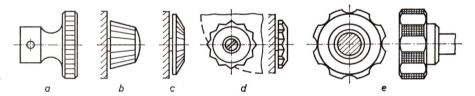

Fig. 1.427. Knobs with knurled rim.

a. Turned metal.　　d. Partially recessed into cabinet.
b. Plastics.　　　　e. Robust, for use as a fastener.
c. Very shallow.

Fig. 1.428. Small knurled rotary knob, used also for push-pull operation.
　a. Spherical.　b. Cylindrical.

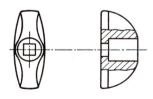

Fig. 1.429. Rotary knob for switch.

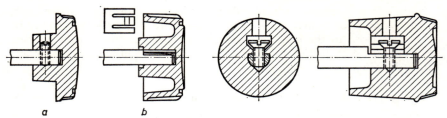

a   b

Fig. 1.430. Methods of fixing knob on control spindle.
a. With grub screw in tapped hole in spindle.
b. With spring clip.

Fig. 1.431. Knob fastened with captive screw.

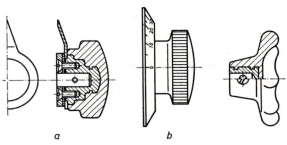

a   b

Fig. 1.432   Fig. 1.433

Fig. 1.432. Knob with dial or pointer.
    a. Cross-section of knob with pointer. b. Side-view of knob with dial.
Fig. 1.433. Handwheel for wrap-around grasp.

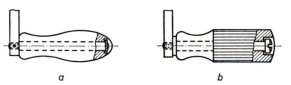

a   b

Fig. 1.434. Fixed crank handles.
    a. With a simple handle.
    b. With a slightly better handle.

A pointer or dial can be fastened to the knob, as shown in Fig. 1.432. Handwheel-type knobs are used to transmit heavy torque: the hand largely encompasses the knob, and the fingers engage depressions in its rim (Fig. 1.433). Larger handwheels are employed for still heavier torque loads, but only occasionally (e.g. in X-ray apparatus).

Cranks are used for prolonged rotary motion. Figure 1.434 shows fixed crank handles and Fig. 1.435 slip-on crank handles.

Turning keys are specially-shaped rotary grips with two opposite leverage faces. The shank is constructed in the same way as that of a slip-on crank. Keys of the plug-on type are used to avoid unauthorized operation of equipment. In such cases, a key with a triangular (or square) socket is preferred and the mating shaft stub is often sunk in a recess as an added precaution (Fig. 1.436). A shaft stub with profiled hole offers less security, because the shaft can still be turned by means of a strip.

Plug-in keys ensure a better appearance, as in the case of a clock, or make the structure less vulnerable, as with roller skates. Other examples are: the keys used on taps and cocks to control the flow of liquid, in door locks and for clockwork toys. Keys that may be left on the shaft stub after use, need to be removed only for occasional dismantling, and normally turn in one direction only (as with an alarm clock), can conveniently be screwed on to a mating thread on the shaft.

Carrying handles are used for picking up or carrying portable apparatus. As with the other handles described, they must have sufficient strength and also fit comfortably into the hand. The types shown in Fig. 1.437 are rounded and deburred or deflashed.

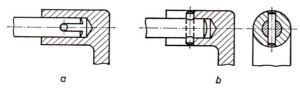

Fig. 1.435. Slip-on crank handles.
a. With pin in hub fitting into slotted shaft.
b. With pin in shaft fitting into slot in hub.

Fig. 1.436. Shaft stub sunk in recess, with triangular end to fit plug-in key.

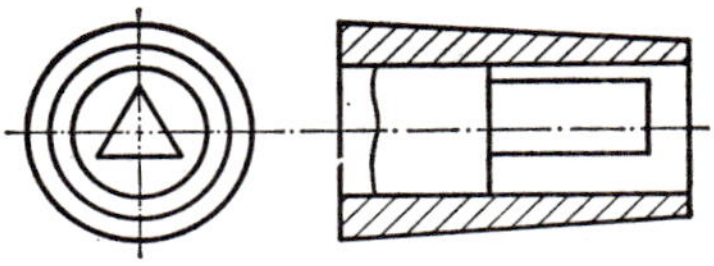

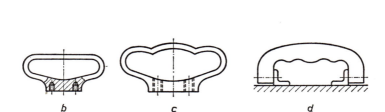

Fig. 1.437. Carrying handles.
a. Metal eyelet or loop for finger, cord or chain.
b. Ditto, to accommodate more than one finger.
c. With rounded plastics grip.
d. Fold-down plastics handle.

REFERENCES

[1] RICHTER, V. VOSS and KOZER, *Bauelemente der Feinmechanik*, Verlag Technik, Berlin, 1954.
[2] G. SCHLEE, *Feinmechanische Bauteile*, Verlag Konrad Wittwer, Stuttgart, 1950.
[3] S. HILDEBRAND, *Feinmechanische Bauelemente*, 1968, Carl Hanser Verlag, Munich.

1.9 Couplings and clutches[1-8, 18]

1.9.1 Introduction

Couplings are used to unite shafts, to transmit rotary motion from one to the other, when the centre-lines of the two shafts coincide or almost do so. The same applies to clutches. Both shafts rotate at the same r.p.m. when the clutch is engaged.

Thus, these devices can be divided into two groups:

Permanent couplings are united by a joint that may be fixed or movable. They include rigid couplings, dog couplings, flexible couplings, universal joints, etc.

Clutches enable the drive, or transmission of motion, to be engaged and disengaged. They may be divided into externally (manually) operated clutches, e.g. of the claw or friction type, and those that function automatically, including slip, centrifugal and one-way clutches.

Mechanical clutches fall into two main groups:

Positive clutches based on the shape of the members, such as, say, interlocking claws, or sets of teeth.

Friction clutches based on the transmission of torque by friction forces generated between the members.

Apart from the purely mechanical clutches, there are others which are electromagnetic, electrodynamic or operated by permanent magnets.

Automatic clutches are classified according to function, e.g.:
● Clutches that disengage beyond (or below) a given angular velocity.
● Clutches that disengage fully or partially when a given torque is exceeded.

Since there are many different versions of each of these types of clutch or coupling, it is beyond the scope of the present chapter to discuss them exhaustively. What follows will deal mainly with the principal characteristics and illustrate current applications in precision engineering.

1.9.2 Rigid couplings

Couplings wherein both shafts are joined without any freedom of movement between them are rare in precision engineering mechanisms. They are sometimes used for shaft extension on members of standard shaft length, with a view to type limitation. The two shafts must be accurately in-line to avoid wobble (Fig. 1.438). Two screws per shaft fastening are better than one, for, if the fit is on the loose side, the shaft will have some freedom of movement, unless fastened in the manner of Fig. 1.438*d*.

In most cases, both shafts will be in double bearings and therefore extremely difficult to align. Imperfect alignment, vibration and deflection can be compensated by means of a coupling with some freedom of movement. No such freedom (or, at any rate, none worth mentioning), exists in the couplings illustrated in Fig. 1.439. The couplings in Fig. 1.440 allow slightly more latitude, particularly in the axial direction. A loose fit between the members improves the radial freedom of movement, but usually also adds to the backlash, especially when the bearing faces are near the centre-line.

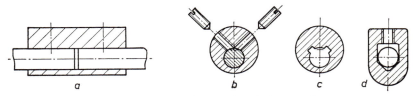

Fig. 1.438. Rigid coupling for shaft extension.

a. Longitudinal section.
b. Construction not recommended. Shaft damage caused by screws makes demounting difficult.
c. Damage to shaft does not affect demounting.
d. Good joint with only one screw per shaft.

Fig. 1.439. Rigid coupling without freedom of movement.

a. Shafts dowelled together direct.
b. Shafts dowelled separately to muff.

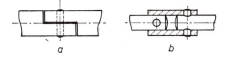

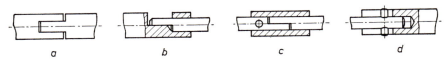

Fig. 1.440. Rigid coupling with only slight freedom of movement (depending on fit), except in axial direction, where there is much greater freedom.

    a. Tongue-and-groove coupling.
    b. Shaft with flat face fits in matching bore of milled shaft.
    c. Shafts with flat faces joined by muff dowelled to one of them.
    d. Shaft with dowel drops into bore of shaft with milled slot.

1.9.3 Dog couplings

To transmit heavier torque loads, reduce backlash and permit axial displacement and errors of radial alignment, a flange coupling or a dog coupling may be selected. Since the flanges of the coupling members merely serve to accommodate the actual coupling parts and the shaft fastening, what in effect are cranks are often used instead of circular flanges, forming a tongue-and-groove type dog coupling. To transmit heavy torque loads the coupling members are often made in the form of lugs or claws engaging a large bearing surface in corresponding openings, as in the case of the claw clutch of Fig. 1.441, where the coupling members are identical. But the use of two claws necessitates strict alignment, which is not necessary in systems with only one claw.

Pins can be used instead of claws for smaller loads. To enable the clutch to be engaged without moving one of the shafts, a spring pin can be employed, as in Fig. 1.442.

A tapered pin instead of a cylindrical one ensures transmission without backlash. Figure 1.443 shows a simple dog coupling which can be provided with an electrically insulated dog or pin.

Dog couplings free from play or backlash, shown in Fig. 1.444, are suitable for rotation in either direction. The spring force must exceed the force to be transmitted by a sufficient margin.

The tongue-and-groove dog coupling is out of angular alignment if the centre-lines of the united shafts are not in line. This coupling is sometimes used to correct the response of a tuning control. The magnitude of the correction can be adjusted by varying the distance between centres of the two (parallel) spindles. See also Volume 1, Section 2.7.2.

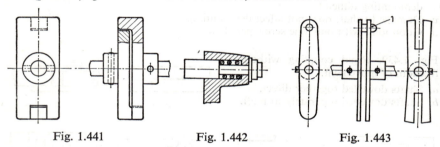

Fig. 1.441   Fig. 1.442   Fig. 1.443

Fig. 1.441. Claw coupling.
Fig. 1.442. Spring pin can engage in coupling flange without moving one of the shafts.
Fig. 1.443. Simple dog coupling, not free from play.
          1 = insulated cap fitted on pin.

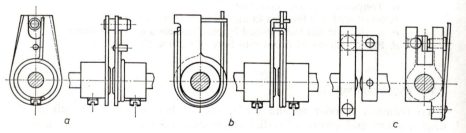

Fig. 1.444. Dog couplings free from play.
   a. Flexural spring eliminates gap between dog and slot.
   b. The two cranks are held together by a hairpin spring.
   c. Dog pressed home by flat spring eliminates play.

1.9.4 Constant velocity (angular truth) couplings for shafts out of alignment

They can be constructed from relatively simple parts. The error introduced by a single dog and slot can be eliminated by using two such couplings, which involves introducing an intermediate shaft with its axis exactly midway between the centre-lines of the two shafts to be coupled. Not that this method is used in practice, however.

A better solution to the problem is the coupling with an intermediate slotted disc (Fig. 1.445). The cross centre-disc (Oldham-type) coupling shown in Fig. 1.445c is suitable for higher torques. Two knob projections on either side of the centre disc slide in mating slots in the discs mounted on the shafts. These slots are always at right-angles. Because the motion is transmitted twice, the backlash is twice that of an ordinary dog coupling.

Fig. 1.446 shows simple devices for coupling with angular truth. A point to remember is that intermediate members of this kind slide and hinge continually, and therefore have to be most resistant to wear and properly lubricated. Hence, they are not recommended.

The angular truth of the aforementioned couplings is readily explained. Since the intermediate member invariably assumes the same position in relation to the shafts, they also remain continually in the same relative position.

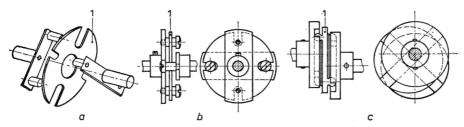

Fig. 1.445. Cross centre-disc (Oldham-type) coupling.

a. Principle of transmission without loss of angular truth.
b. Lightweight construction with shouldered screws on cranks.
c. Heavy-duty construction with two knobs, projecting from either side of centre disc, engaging slots in flanges mounted on shafts.
   1 = cross centre disc.

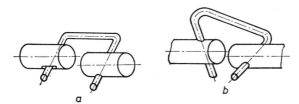

Fig. 1.446. Simple angular-truth coupling.

a. U-shaped bracket able to slide sufficiently in suitable traverse holes in shaft.
b. As a, but bracket limbs cross at right-angles.

1.9.5 *Flexible couplings*

Flexible disc couplings are used to cushion vibration and jolting in drives, e.g. motor drives. Two examples are illustrated in Fig. 1.447. The motion is transmitted via a disc or one or two bushes of flexible material, made in one piece of rubber or built-up of leather discs. Other and heavier versions of the flexible disc coupling also occur. For lighter torque loads, a highly simplified version can be used, with a rubber hose as the flexible coupling member (Fig. 1.448). A piece of steel wire (suitable for rotation in one direction only) or a helical spring (Fig. 1.449) can be employed, instead of the rubber hose. These couplings can unite shafts when they are out of angular alignment and/or when the relative angle or position of the shafts changes in operation.

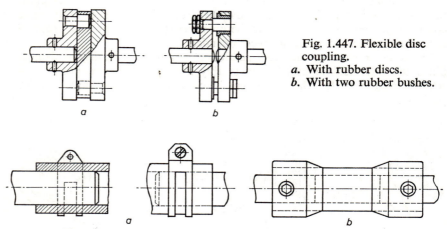

Fig. 1.447. Flexible disc coupling.
a. With rubber discs.
b. With two rubber bushes.

Fig. 1.448. Flexible coupling with rubber hose as coupling member.
a. Hose fastened to shafts by hose clips.
b. Hose, with metal bushes vulcanized to it, fastened to shafts by screws.

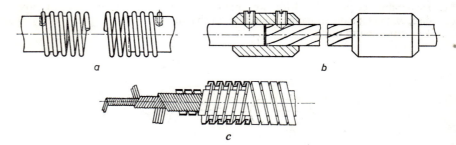

Fig. 1.449. Coupling with flexible element.
a. Helical spring, one end of which engages in transverse hole in shaft.
b. Steel wire clamped in bushes.
c. Flexible shaft composed of helical springs with steel sheath.

An improved version of the helical spring coupling is the flexible shaft composed of multi-turn helical springs wound alternately clockwise and anticlockwise (Fig. 1.449c). For better handling and for protection the shaft is enveloped in a steel sheath.

1.9.6 Universal joints[10, 11]

These couplings (or joints) are used to unite two shafts making an angle of 135° to 180° and crossing at the coupling. Apart from one or two minor artifices, to be described later, angular compensation (truth) is not involved.

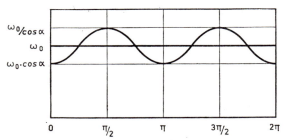

Fig. 1.450. Single universal joint. Angular velocity curve of driven shaft compared with angular velocity of driving shaft (ω_o), plotted against angle traversed by this shaft. For α see Fig. 1.451.

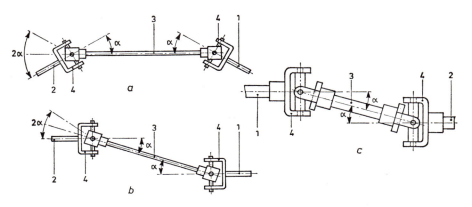

Fig. 1.451. Arrangement of double universal joint providing angular compensation. All three shafts in the same plane.

a. Driving and driven shaft at same angle (α) to intermediate shaft.
b. Driving and driven shaft parallel.
c. Correct relative position of coupling members on intermediate shaft for transmission without angular misalignment.

$\qquad$ 1 = driving shaft. $\qquad$ 3 = intermediate shaft.
$\qquad$ 2 = driven shaft. $\qquad$ 4 = universal joint.

When the driving shaft rotates at an angular velocity ω_0, the variation in angular velocity of the driven shaft, linked to the driving shaft by a universal joint, is limited by the factors $\omega_0/\cos\alpha$ and $\omega_0.\cos\alpha$, as plotted for a full revolution in Fig. 1.450 (for α see Fig. 1.451). This also establishes the maximum ratio of the instantaneous velocities of the two shafts, from which it will be seen that the deviation may be substantial. This difference can be eliminated completely by employing an intermediate shaft, so that the coupling is effected by two universal joints. Thus, the positive deviation of one coupling is compensated by an equivalent negative deviation of the other. Fig. 1.451 illustrates two possibilities involving shafts in the same plane. Note that the coupling members are so positioned on the intermediate shaft as to ensure longitudinal symmetry of this shaft. As well as in the position indicated, there is no loss of angular truth when the shaft moves to any other position on the conical shell within an apex angle of 2α (Fig. 1.451). See also Volume 1, Section 2.7.4.

Universal joints can be grouped under the following headings:

(a) Hooke's joints (cross trunnion) or cardan joints.
(b) Ball universal joints.
(c) Elastic ring and diaphragm couplings.

(a) *Hooke's joints (or couplings) and cardan joints*

These are couplings with a central coupling member between the shafts. The coupling member is cross-shaped in the Hooke's joint and ring-shaped in the cardan joint, and has two pivot axes crossing at right-angles at the point of intersection of the shafts (Fig. 1.452).

Fig. 1.453a is an example of a Hooke-type joint with a cylindrical coupling member, or spider, which has a pair of screws as pivots. For lighter duty a a ball with a pair of grooves turned in it can be used as the coupling member, as shown in Fig. 1.453b. The ends of the four prongs are curved slightly round the ball to prevent any axial displacement.

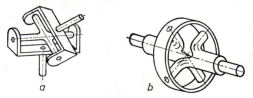

Fig. 1.452.   *a.* Hooke's joint.   *b.* Cardan joint.

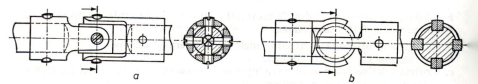

Fig. 1.453. Hooke's joints.
a. Cylindrical centre-piece as spider.
b. Ball with slots at right-angles, instead of a cross.

(b) *Ball universal joints*

Are basically ball joints, in which it is practically impossible for the pivots to twist relative to one another in the principal direction of rotation. The coupling shown in Fig. 1.454 is suitable only for low-torque transmission, because of the limited area of contact between pin and ball, the sliding motion of the ball relative to the pin, and the limited radius of action of the tangential force. Simpler and cheaper structures are given in Fig. 1.455. Rational integration of the elements shown in Fig. 1.451 leads to the design of Fig. 1.456. The coupling of Fig. 1.457 has equal angular truth if the point of contact of the two embedded drivers (or dogs) is within the angle of intersection β between the two shafts. The force is transmitted via a ball.

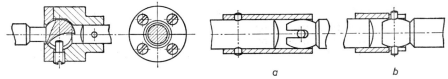

Fig. 1.454. Ball universal joint for small torque loads.

Fig. 1.455. Ball universal joint of simple design.
a. Spherical fork rotates and slides in socket and round impressed pin.
b. Opposite version of *a.*

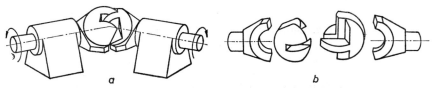

Fig. 1.456. Double ball universal joint for angular truth.

a. Assembled.   *b.* Exploded view.

Fig. 1.457. Ball universal joint for angular truth, with two dogs to transmit the motion via pressure contact.
1 = dog.   2 = ball.

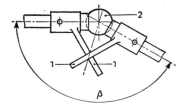

(c) *Elastic ring and diaphragm couplings*

Elastic ring couplings are often used in precision engineering for shafts that are marginally misaligned and have to transmit only a low torque (Fig. 1.458). They are flexible, rather than universal, their advantage over true universal joints being their freedom from backlash, whilst still permitting minor axial displacements. If necessary, the ring can be punched out of insulating material. Diaphragm couplings (Fig. 1.459), suitable for transmitting heavier torques, are less flexible in operation and more expensive than the elastic ring couplings.

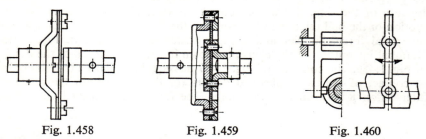

Fig. 1.458   Fig. 1.459   Fig. 1.460

Fig. 1.458. Elastic ring coupling.
Fig. 1.459. Diaphragm coupling.
Fig. 1.460. Coupling operated by a sliding bush, driven by a forked crank.

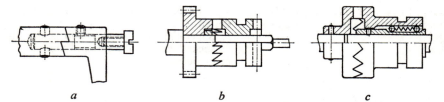

a   b   c

Fig. 1.461. Claw clutch with bevelled teeth.

a. Single bevel on hand-operated crank.
b. Single bevel on mechanism to set hands of clock.
c. Double bevel, suitable for rotation in both directions. Practical version is suitable for continual engagement and disengagement.

1.9.7 Claw clutches

Most claw clutches can be engaged only with the shafts stationary in a given relative position, so that the teeth line up with the recesses designed to receive them. The clutches can be disengaged with the shafts rotating or stationary. The sliding member of the coupling can be actuated by a sliding shaft, driven by a forked crank (Fig. 1.460). So that the clutch can be engaged in any position, the mating faces of the teeth are bevelled. With a single bevel (or taper), forces can be transmitted only in one direction (Fig. 1.461*a* and *b*): for rotation in either direction, the teeth are tapered on both sides (Fig. 1.461*c*). This clutch is held in engagement by a strong spring force that must be overcome to disengage the clutch: the sliding bush moves on balls. To enable the clutch to stand up to continual engagement and disengagement, the teeth are made of hardened and tempered (or locally hardened) tool steel.

1.9.8 Friction clutches[9, 15, 17]

These clutches are used for the smooth (jolt-free) engagement and disengagement that claw clutches cannot provide. The short clutch travel is often an advantage. The surface pressure required to set up the frictional force is derived from muscular strength, spring pressure, centrifugal force or electromagnetic power. To obtain a maximum of friction with a minimum of torque, the following are employed: plates (or discs) or multiple plates; a friction surface in the form of a cone shell (cone clutch).

In plate or multiple plate clutches (Fig. 1.462), several friction surfaces are pressed together by one and the same coupling force, thereby multiplying the frictional forces generated and permitting a compact structure. In the cone clutch, the frictional forces are increased by the slope of the conical members. Referring to Fig. 1.463:

$$N = \frac{F}{\sin \alpha}; \quad T = \mu N = \frac{\mu F}{\sin \alpha} \quad \text{and} \quad M = T.r = \frac{\mu F.r}{\sin \alpha}$$

where F is the axial actuating force;
   α is half the apex angle of the cone;
   N is the force normal to the friction surface;
   T is the tangential frictional force;
   μ is the coefficient of friction between the friction surfaces;
   M is the torque to be transmitted.

To prevent sticking, and to keep the force required to disengage the clutch as small as possible, the angle α should be larger than the angle of friction.

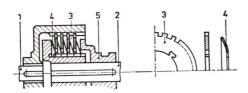

Fig. 1.462. Multiple plate clutch.

1 = driving shaft.   4 = inner plate.
2 = driven shaft.   5 = sliding bush.
3 = outer plate.

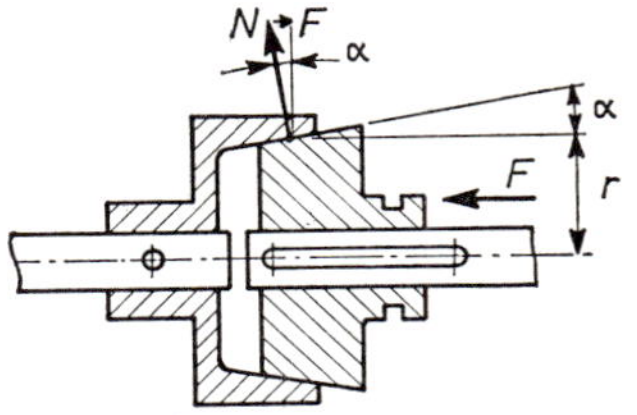

Fig. 1.463. Forces acting on a cone clutch.

Steel is usually chosen as the friction material on the one hand and, amongst other things, phosphor bronze, fibre, lignum vitae or plastics on the other hand. Rubber is also used sometimes in precision engineering for flat friction discs in clutches that transmit a relatively high torque. It has the advantage of a high coefficient of friction, but suffers from the following drawbacks:

● Stick-slip due to the variation in the coefficient of friction with speed.
● Undesired elasticity in clutches with relatively thick friction discs when motion is transmitted.
● Possible sticking on disengagement.

Slip couplings usually serve to limit the torque transmitted (maximum torque). Couplings of this type unite two shafts, in separate bearings, or concentric members, such as a shaft and a gear or ratchet wheel, a shaft and a bush or a bush and a pointer, and are often used to protect a fragile mechanism (Figs. 1.464 and 1.465). The frictional force is generated by an elastic member or by a separate star-shaped spring, bowed (saddle) spring or helical spring. The type of maximum torque coupling with a braking spring in a drum (Fig. 1.466) has been discussed, together with the relevant calculations, in Volume 1, Section 2.3.7.

Fig. 1.467 shows couplings in which the frictional force is increased by a conical friction surface or by double friction plates.

A problem associated with such cone friction couplings is the centering of the two conical members in relation to one another, which requires very accurate bearing alignment. The solution is to mount the shafts in bearings one around the other. If a helical spring used to press a friction plate home is not stiff enough, it may "wind up" in the direction of rotation. In critical cases this can be remedied by interposing an axial thrust ball bearing.

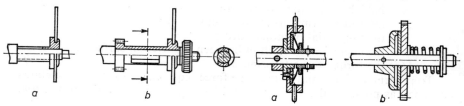

a   b   a   b

Fig. 1.464. Simple slip coupling.

a. Spring pointer bush on shaft.

b. Spring pointer bush with thicker walls and fitted with a gear wheel.

Fig. 1.465. Slip coupling.

a. Bowed spring keeps pin wheel in frictional engagement with shaft, as long as the torque to be transmitted does not become excessive.

b. Helical spring keeps gear wheel in frictional engagement with shaft, if the torque to be transmitted does not increase unduly.

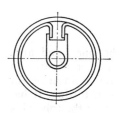

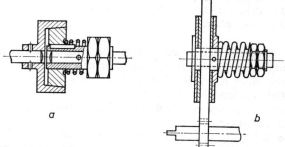

a   b

Fig. 1.466. Overload coupling with braking spring in drum.

Fig. 1.467. Adjustable friction clutch for heavier torque load.

a. Cone clutch.   *b.* With double friction plates.

When the slip torque has to be within fairly exact limits, it should be borne in mind that the spread in the force normal to the friction surface, the coefficient of friction and the radius of friction, all contribute to the spread in the slip torque. The force of a spring can be adjusted, or a flexible spring chosen instead. The size of the friction radius can be established fairly accurately in the design (Fig. 1.465b). Fibre is often used as the friction material. The friction disc can be provided with compartments to store a special silicone oil which lubricates, and continues to lubricate, the friction surface, thereby ensuring that the coefficient of friction does not deviate very much, even after several years of service, and is not affected by occasional drops of lubricating oil spilled on the disc. Dry lubricants, such as molybdenum disulphite (MoS_2) have a beneficial effect.

If the friction disc starts to slip because the torque is beyond the maximum for which it is designed or adjusted, the maximum torque remains perceptible at the driving side. Nevertheless, the perceptible torque will diminish by reason of the difference in friction when stationary and in motion.

Leather has special characteristics, however, provided the dressed side of the oiled leather is allowed to slip against the other friction member. The coefficient of friction, and therefore also the torque, increases with the slip speed. A coupling with this friction material runs smoothly and does not snatch. A friction material having a coefficient of friction that does not depend on the slip speed is the thermoplastic Teflon.

Reference values of friction coefficients for the friction materials used in couplings are listed in Table 1.1:

TABLE 1.1

Reference values of friction coefficients for friction materials

Friction face	Backing	Coefficient of friction		Max. temp. (°C)	Max. surface pressure (N/cm²)
		Lubricated	Dry		
Phosphor bronze	Chr. plated hard steel	0·03–0·05	—	260	100
Hard steel	Chr. plated hard steel	0·03	—	260	140
Lignum vitae	Steel	0·16	0·2–0·35	200	40–60
Leather	Steel	0·12–0·15	0·3–0·5	90	7–28
Leather	Oak	—	0·3–0·5	90	—
Fibre	Steel	—	0·3–0·5	90	7–28
Cork	Steel	0·15–0·25	0·3–0·5	90	5–10
Felt	Steel	0·18	0·22	135	3·5–70
Bakelite Synthetic resin bonded paper Synthetic resin bonded fabric	Steel	0·1–0·15	0·25	200	70
Teflon	Steel	—	0·05	300	—

1.9.9 *Overriding clutches*

In these clutches the drive is transmitted between the coupling members by overriders (sprags, balls, etc.), which may be free from backlash. As seen in Section 1.6.5, a frictional resistance of $F(\tan \alpha + \mu)$ can thus be transmitted instead of $F.\mu$, as in the case of a flat friction clutch, where F is the axial coupling force, α is the angle of inclination of the overrider face and μ is the coefficient of friction. An overrider clutch can be used to best advantage where a relatively high torque has to be transmitted with a small coupling force (F), and the clatter associated with slip is not a hindrance, or may even be useful as an indication that slipping is taking place.

Fig. 1.468 shows two ball-overrider clutches, one with radial, and the other with axial action. They are not really designed for frequent freewheeling, owing to the wear involved. Bearing surfaces of members running against one another must be hardened. These clutches are usually employed as maximum torque governors.

A special version is the overriding clutch, permitting rotation in one direction only, of the type used for micrometer screw adjustment to apply measuring pressure within given limits (Fig. 1.469).

Two simple overload (overrunning) clutches are shown in Fig. 1.470.

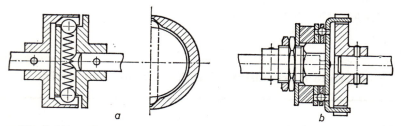

Fig. 1.468. Ball-overrider clutch. *a.* Radial-action. *b.* Axial-action.

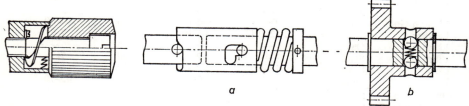

Fig. 1.469. Overriding clutch in micrometer.

Fig. 1.470. Simple overrunning clutches.
a. Overload pushes sleeve to the right, disengaging the clutch.
b. Overload pushes balls inwards, disengaging the clutch.

1.9.10 Magnetic and electric clutches[12, 16]

(a) Introduction

These clutches fulfil the same functions as mechanical clutches, but use magnetic fields. The action of electric (electromagnetic) clutches, unlike the mechanisms involving permanent magnets, can be controlled. The current in the field-coil can be switched and regulated to engage (and disengage) the clutch, and also control its operation in various ways.

The electrical equipment not only makes the clutch fast and easy to operate, but also enables it to be controlled from more than one station and remotely: frequent engagement and disengagement, and fast response, can be achieved economically.

Some clutches of this type are eminently suitable for use as overload (or overrunning) clutches or as speed governors.

Electrical (electromagnetic) clutches can be grouped under the following headings (Fig. 1.471):

- Single-plate or multi-plate friction clutches.
- Claw clutches with teeth tapered on both sides.
- Hysteresis clutches.
- Eddy current clutches.

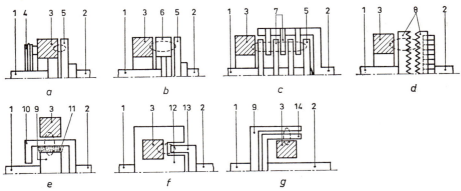

Fig. 1.471. Principles of electric clutches.

a. Electromagnetic clutch with friction plate (flat armature clutch) and rotating field.
b. As a, but with stationary field.
c. Electromagnetic clutch with friction plates and stationary field.
d. Electromagnetic claw clutch with teeth tapered on both sides.
e. Electromagnetic clutch with ferromagnetic particles.
f. Hysteresis clutch.
g. Eddy current clutch.

1 = driving shaft.　　2 = driven shaft.　　3 = field winding.　　4 = slip rings. 5 = armature.　　6 = rotor.　　7 = friction plates.　　8 = clutch claw with tapered teeth.　　9 = driving rotor.　　10 = driven rotor.　　11 = ferromagnetic particles. 12 = multi-pole rotor.　　13 = bush armature.　　14 = copper bush.

In addition, there are clutches which, although purely mechanical in principle, are engaged and disengaged indirectly by electrical means. This group will not be discussed.

Some clutches operate with a rotating magnetic field, others with a stationary magnetic field. This is a matter of magnet design.

Rotary field versions have a rotating field winding, energized via brushes and sliprings: in stationary field clutches, the field winding is fixed. A rotating field clutch is usually a two-part functional element with the field winding at the driving end. The clutch with a fixed field coil is a three-part element having a rotary driving member and a fixed coil housing. It does not require brushes and sliprings, but does call for extra bearing points and better alignment.

The use of a direct-current supply, usually 90 V, is considered a drawback. Miniature clutches for aircraft, etc., are designed to operate at 24 V; for cars and machinery, 6 V or 12 V is usual. On no account should anti-friction (rolling motion) bearings be placed within the magnetic (leakage) field.

The development of electric clutches is still being vigorously pursued, the trend of research being towards still shorter response times (0·3 milliseconds), which it is hoped to achieve through principles not involving the setting-up of a magnetic field (electrostatic and electrostrictive clutches) or making use of an existing field which has only to be engaged. Such clutches are intended for very fast tape recorders, computers and telephone systems.

(b) *Magnetic clutches*[13]

One special type of overriding clutch is the magnetic clutch, in which the motion of the driving shaft is transmitted from a liquid-tight chamber to a pointer spindle outside, without piercing the chamber wall (Fig. 1.472a). The armature on the pointer spindle follows the two permanent magnets, which rotate with the driving shaft. The dividing wall is made of non-ferrous material (and preferably a non-metal) to avoid eddy current losses.

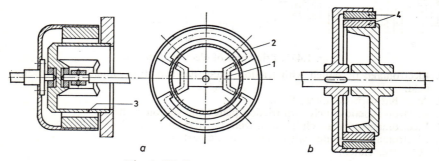

Fig. 1.472. Permanent magnet clutch.

a. Operates through closed wall.   b. Overrunning clutch with annular magnets.
    1 = magnet on driving shaft.   3 = closed wall.
    2 = armature on pointer spindle.   4 = annular multipole magnets.

Magnetic clutches can be used as overrunning clutches (Fig. 1.472*b*). Annular permanent magnets are cemented to the clutch discs. The annular magnets are so magnetized as to provide equal numbers of pole pairs on both rings. Thus, one disc drives the other until a given maximum torque is reached, whereupon the clutch disengages. The residual torque remains low until the relative speed of the two discs has fallen enough for the clutch to re-engage.

(c) *Electromagnetic friction clutches*

They may be single-plate or multi-plate clutches. The electromagnet of the *plate clutch* or flat armature clutch attracts the segmented armature against the ring of friction material fastened to the coil housing (Fig. 1.473*a*). The field coil of this version rotates with the clutch. Other versions with a stationary field coil (Fig. 1.471*b*) are used too. The magnetic flux then passes through the rotor to the armature and back. This friction clutch is used for engagement and disengagement (short switching times), but can also be used as a continuous slip clutch, although this is not recommended, in view of the extra heat generated. Smaller models are employed in instruments and in aircraft equipment.

The electromagnetic *multi-plate clutch* can transmit a very much greater torque than the single-plate model. It is also designed for on–off operation, and can be equipped with a rotating field coil or with a stationary one, the latter version being the more common of the two (Fig. 1.473*b*). It is capable of transmitting a substantial torque load within a small space. Most clutches of this type run in an oil bath, and are suitable for machine tools and control systems.

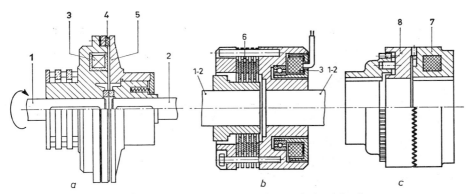

Fig. 1.473. Electromagnetic friction and claw clutches.

a. Friction plate clutch.   *c.* Claw clutch with tapered teeth.
b. Multi-plate friction clutch.
1 = driving shaft.   2 = driven shaft.   3 = field coil.   4 = friction ring.   5 = flat armature.   6 = multi-plate friction unit.   7 = toothed magnet member.
8 = toothed clutch disc.

(d) *Electromagnetic claw clutches*

Are able to transmit high torque loads but can only be engaged at low speed. The clutch (Fig. 1.473c) consists of the toothed magnet member with field coil and the toothed clutch disc, of which the stationary field coil version is the most common.

The clutch is used mostly in numerical control systems.

(e) *Electromagnetic clutches with ferromagnetic particles*[14]

They consist of a driving rotor enclosed in (and concentric with) a driven rotor. These two rotating members are separated by an air gap filled with ferromagnetic particles (Fig. 1.474). Until the field coil is energized the driving rotor freewheels. When energized (short response time), the iron particles quickly form a solid mass, capable of transmitting forces, the value of which does not depend on the relative speeds of the driving and driven rotors. The torque transmitted is thus constant and independent of the speed of the driven shaft. Continuous slipping is possible, given due attention to cooling.

This clutch was originally designed to operate with iron particles in oil, but the present trend is towards an air gap filled solely with ferromagnetic particles.

Electromagnetic clutches are used in machines and instruments, particularly where a substantial mass moment of inertia has to be overcome in engaging the clutch.

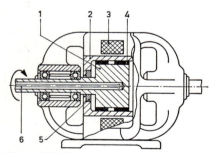

Fig. 1.474. Electromagnetic clutch with ferromagnetic particles.
1 = driving rotor.
2 = driven rotor.
3 = field coil.
4 = ferromagnetic particles.
5 = seal.
6 = cooling ducts.

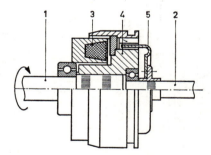

Fig. 1.475. Hysteresis clutch.
1 = driving shaft.
2 = driven shaft.
3 = field coil.
4 = rotor with several poles.
5 = bush armature of hard magnetic material.

(f) *Hysteresis clutches*

The action of these clutches is based on the hysteresis effect produced by the pole reversal of hard magnetic material. The driving member is a multi-pole rotor, to set up the desired alternating field (Fig. 1.475). The driven member is a bush armature of hard magnetic material which fits in the annular slot in the rotor. In this version, the field does not rotate. When the field coil is energized, magnetic flux passes through rotor and armature. The poles in the rotor concentrate the flux in such a way as to induce opposite poles in the armature, which rotates with the motor. The rotating rotor cuts through the constant focused flux of the field coil, which causes the magnetic field in the rotor to change. Because of hysteresis in the armature, the flux change in it cannot keep pace with that induced by the rotor. The armature therefore lags behind the rotor, whereby torque is transmitted.

To suppress possible eddy currents, the armature is laminated. The clutch can be used with any amount of slip, consistent with assured heat dissipation, whilst the torque remains constant. The size of this torque can be regulated by varying the energizing current of the field coil. Clutches in which the field coil is replaced by permanent magnets have a fixed torque.

By virtue of smooth starting and constant torque at varying speeds, this clutch is suitable for instruments, servomechanisms, aerial drives and take-up systems for tape or wire (see Volume 5, Chapter 5).

(g) *Eddy current clutches*

Cannot function without a certain amount of slip. The torque increases with the slip up to a certain critical value. The clutch in Fig. 1.476 has a stationary field coil, generating magnetic flux which circulates through the driven multi-pole rotor and the stationary housing. Between rotor and housing there is an air-gap containing a copper bush fixed to the driven shaft. Any difference in speed between rotor and bush sets up eddy currents in the bush, whereby torque is transmitted from the driving shaft to the driven shaft. The size of the torque depends on the energizing current of the field coil and the amount of slip. A tachometer on the driven shaft measures the speed of rotation.

The energy dissipated by the eddy current must be channelled off, which means that powerful clutches of this type have to be air-cooled or water-cooled. The clutches should preferably be operated with only a small amount of slip. Used in combination with an electronic control and one or two additional measuring instruments, this clutch operates as a speed governor. As such, it has many possible uses as a source of: constant angular velocity; constant take-up winding speed of, say, a wire; constant wire tension in coil winding; defined smooth, positive starting.

Thus the device is employed as a controlled speed drive rather than an actual clutch for winding processes in paper-making machines, textile machinery, etc. (see Volume 5, Chapter 5).

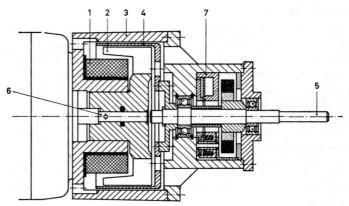

Fig. 1.476. Eddy current clutch.

1 = field coil.   2 = multi-pole rotor.   3 = housing.   4 = copper bush.
5 = driven shaft.   6 = driving shaft.   7 = tachometer.

1.9.11 Centrifugal clutches

Operate automatically, centrifugal force being used to engage them. Almost all the existing types are friction clutches, classified as free-shoe (direct action) and spring-shoe (indirect action) (Fig. 1.477).

In the free-shoe type, the normal force required to generate friction is derived from the centrifugal forces of radially sliding masses, and the transmissible torque is proportional to the square of the angular velocity of the clutch. It functions as a slip clutch until the speed of rotation builds-up enough to deliver the centrifugal force required for friction. Fig. 1.477c shows a practical free-shoe centrifugal clutch with pivoted shoes. Such clutches are used when a shaft has to be set in motion very smoothly.

In the spring-shoe centrifugal clutch, the normal force required to generate friction is derived from a system of springs. As the angular velocity builds up, the spring force is gradually overcome by the centrifugal forces of radially sliding masses, whereby the frictional forces are reduced and may approach zero. This type of clutch is used for automatic speed control.

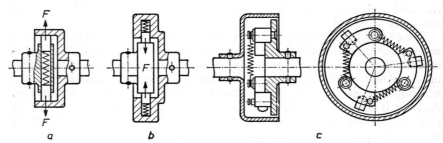

Fig. 1.477. Centrifugal clutches.

a. Free-shoe type.   *c.* Free-shoe, with three pivoted shoes.
b. Spring-shoe type.

1.9.12 One-way clutches

The members of a one-way clutch engage only in one direction of rotation and freewheel in the other direction. Toothed ratchets, friction ratchets, locking elements or a torsion spring are used as drivers. Locking elements and friction ratchets supply a considerable centripetal force. One-way clutches occur in clocks and bicycles and many different versions are frequently encountered in machines used for mechanized processes.

Fig. 1.478 shows ratchet mechanisms, whilst clutches with locking elements (balls or rollers) are illustrated in Fig. 1.479.

The freewheel clutch in Fig. 1.480 is used in bicycles. To minimize the angle of backlash that occurs before the clutch engages, the number of teeth is different from the number of locking elements. In this case, there are three pawls and seven teeth. Only one pawl bears at a time.

A clutch with many locking elements functioning as friction pawls (or sprags) in a strong and compact structure is shown in Fig. 1.481. The locking elements are threaded on an elastic link, so that they are ready for instant engagement. There is virtually no backlash angle.

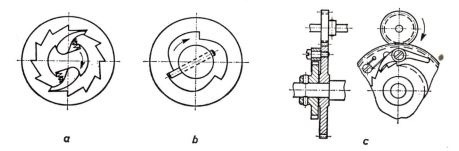

Fig. 1.478. Ratchet clutch.
a. With two simple rotating pawls.
b. With sliding pin as pawl (in a garden mower)
c. Freewheel clutch (in a clock).

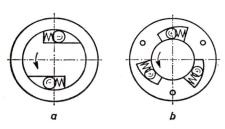

Fig. 1.479. Unidirectional clutch with locking elements.
 a. In rotor.   b. In housing.

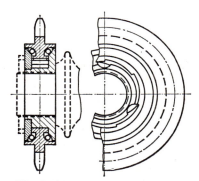

Fig. 1.480. Freewheel clutch.

A typical construction of thermoplastic synthetic resin (Fig. 1.482) operates with numerous sprags in a cylindrical housing, and is suitable for light loads. The resilience involved in engagement may be a drawback or an advantage in practical application. The clutch in Fig. 1.483 is also unidirectional by reason of the shape of its teeth (see also Figs. 1.461*b* and 1.469).

Mechanisms with a helical spring (Fig. 1.484) are suitable for small torques. These torsion spring constructions are described more fully in Volume 1, Section 2.3.7. The shaft is gripped when the direction of rotation is acting to wind up the spring. These clutches are critical, and a certain angle of backlash occurs before the spring engages.

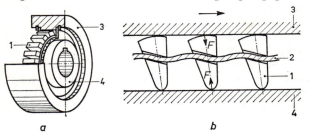

Fig. 1.481. Strong and compact unidirectional clutch with friction pawls (sprags).
a. Part of outer ring ommitted.
b. Enlarged detail of sprags shown schematically.

    1 = sprags.   3 = outer ring.
    2 = elastic link. 4 = inner ring.

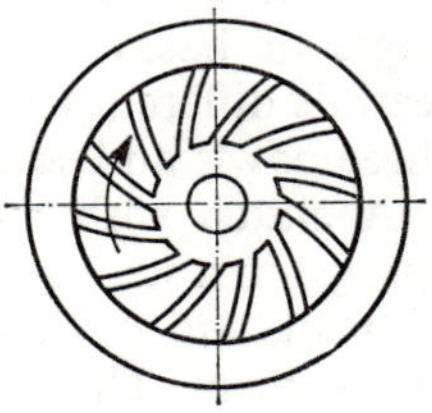

Fig. 1.482. Simple unidirectional clutch with rotor and sprags united in one moulding.

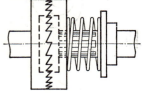

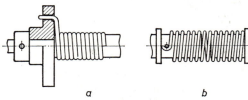

Fig. 1.483. Claw clutch with angled teeth, as unidirectional clutch.

Fig. 1.484. Unidirectional clutch with torsion spring.
a. Spring and disc are driven in one direction by shaft.
b. Spring couples in one direction of rotation.

REFERENCES

[1] RICHTER, V. VOSS and KOZER, *Bauelemente der Feinwerkmechanik*, Verlag Technik Berlin, 1954.

[2] K. HAIN, *Die Feinwerktechnik*, Fachbuchverlag Dr. Pfanneberg, Giessen, 1953.

[3] G. SCHLEE, *Feinmechanische Bauteile*, Verlag Konrad Wittwer, Stuttgart, 1950.

[4] D. C. GREENWOOD, *Product Engineering Design Manual*, McGraw-Hill, New York/Toronto/London, 1959.

[5] K. RABE, *Grundlagen Feinmechanischer Konstruktionen*, Anton Ziemsen Verlag, Wittenberg/Lutherstadt, 1942.

[6] F. WOLF, *Lagerungen, Geradführungen und Kupplungen in der Feinwerktechnik*, Deutscher Fachzeitschriften und Fachbuchverlag, Stuttgart, 1956.

[7] *Mechanical Drives Reference Issue*, Machine Design, 1967, September 21.

[8] K. CENTMAIER and H. EDER, *Zur systematik der feinwerktechnischen Kupplungen*, Feinwerktechnik, 1964, 11, p. 447.

[9] C. F. WIEBUSCH, *Dial Clutch of the Spring type*, The Bell System Techn. Journal, 1938, p. 724.

[10] A. H. RZEPPA, *Universal Joint Drives*, Machine Design, 1953, April, p. 162.

[11] W. REUTHE, *Die Bewegungsverhältnisse bei Kreuzgelenkantrieben*, Konstruktion, 1950, p. 305.

[12] J. F. PECH, *Electric Clutches, Mechanical Drives Reference Issue*, Machine Design, 1967, September 21, p. 46.

[13] G. MUTZ, *Grundlagen für den Einsatz von Magnetkupplungen in der Feinwerktechnik*, Feinwerktechnik, 1967, 8, p. 374.

[14] W. I. TURTSCHENKOW, *Neues Nachlaufsteuerschema mit elektromagnetischen Pulverkupplungen*, Feinwerktechnik, 1968, 1, p. 29.

[15] K. ROTH, *Reibkupplungen in der Feinwerktechnik*, Feinwerktechnik, 1961, 8, p. 285.

[16] N. P. CHIRONIS, *Machine Devices and Instrumentation*, McGraw-Hill, New York/London, 1966.

[17] A. GEMANT, *Frictional Phenomena*, Chemical Publishing Co., Brooklyn, U.S.A., 1950.

[18] S. HILDEBRAND, *Feinmechanische Bauelemente*, 1968, Carl Hanser Verlag, Munich.

1.10 Transmission by tensile or compressive forces[1 – 9]

Tensile forces can be transmitted over relatively long distances by means of cords, belts and chains. Switches and indicators (such as recorders with styluses and pointers, etc.) often contain a drive of this kind.

The direction in which the tensile pull is transmitted can be changed where necessary by means of a guide pin, roller, pulley or sprocket. In the transmission of linear motion, translation can be converted into rotation, and vice versa (Fig. 1.485). Two rotating discs can be linked by such a sliding member. The link may be endless or have its ends fastened to the members thus coupled. In the latter case, the angle of rotation is limited, whilst the transmission ratio may be variable (Fig. 1.486). Another possibility

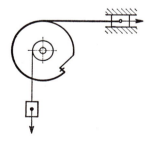

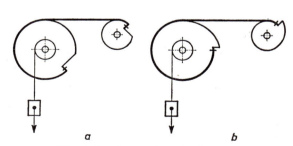

a   b

Fig. 1.485. Conversion of translation into rotation, or vice versa.

Fig. 1.486. Transmission of rotation.
a. Constant transmission ratio.
b. Varying transmission ratio.

is a prestressed cable passing through a flexible sleeve, thereby exerting tensile pull regardless of any change in direction or position of the sleeve inlet relative to its outlet (Bowden cable) (see Section 1.4.2).

Cord made of hemp or cotton (or sometimes of silk, catgut or nylon) is usually employed for drives in radio or T.V. sets and other apparatus (see Section 1.11.6).

Loops of the kind shown in Fig. 1.487a, readily obtained with a crimped metal bush, are used for attachment. Sometimes, a blob of plastics, e.g. polycarbonate, is moulded on to the cord instead (Fig. 1.487b). A tensioning device must be provided to compensate for possible stretching or shrinking of the cord material. Large-scale batch production also makes use of the existing tensioning device to compensate for any spread in the size of the structural members, upon which the length of the cord is based.

Where two linked members are required to move synchronously, they should be arranged in balance, or symmetrically, as regards cord length and tension (Figs. 1.488a and b). Tensioning is provided by a helical spring accommodated in a drum or on the pointer, so as approximately to equalize any elongation and shrinkage in the linkage members. Or a metal cable subjected to very little length variation is used for the span between the driving drum and pointer, and cord with a tensioner for the other span (Fig. 1.488c). In this case synchronism is ensured by the metal cable. The last system is better than the two other methods, in which the pull of the spring is transmitted with friction to the span or side. Where the spring has to keep the side tension within reasonable limits, despite any length variations, the friction encountered by the side on entering the drum, or on meeting the spring on the

Fig. 1.487. Loop formed on cord with the aid of:

a. Crimped metal bush.   *b.* Moulded blob of plastics.

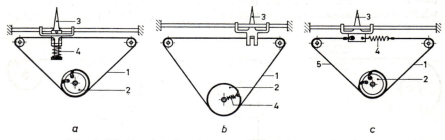

Fig. 1.488. Synchronism between driving drum and pointer.

a. Cord arranged symmetrically with tensioning spring on pointer.
b. Cord arranged symmetrically with tensioning spring in drum.
c. One side consisting of steel cable subjected to very little length variation, the other side using a spring and possibly a driving roller.

1 = cord.   2 = driving drum.   3 = pointer.   4 = spring.   5 = metal cable.

pointer, may be an obstruction. Appreciable variation of side tension has an adverse effect on smooth running, particularly when the drive is imparted by winding one or two turns of the cord round a driving roller (friction roller) (see Section 1.11.6).

Instead of using a separate spring, there is another method of tensioning, shown in Fig. 1.488, obtained by taking the cord round a roller, which exerts a pull on the transmission link. If the link itself is elastic (for example, a rubber belt or a cable in the form of a helical spring), there is no need for a separate tensioner. In any case, it is necessary to make sure that the backlash in the drive is acceptable. The side tension also depends to some extent on the direction of rotation. Variation of this tension may cause the tensioner to throb or pulsate and cause backlash. In Fig. 1.488a, this pulsation is counteracted by the aforementioned friction. It ensures that the device does not keep exactly in step with the tension in the span, although this was also mentioned as a drawback.

Plastics are used for flat belts (see Section 1.11.6) and for endless toothed or cross-ridged belts (timing belts) that are reinforced (Figs. 1.489a and b). High-grade timing belts contain cord of glass fibre wound in a continuous and unbroken spiral, or steel cable made up of thin lengths of wire. The serrations of the chloroprene (Neoprene) body are sheathed in polyamide fabric (nylon). A cord with beads of plastic moulded on to it at intervals is better for some purposes than a metal chain (Fig. 1.489c).

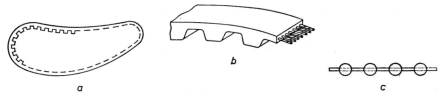

Fig. 1.489. Use of plastics belt.
a. Endless timing belt.
b. Endless timing belt: reinforced and with nylon-sheathed serrations.
c. Cord with beads of plastic moulded on to it.

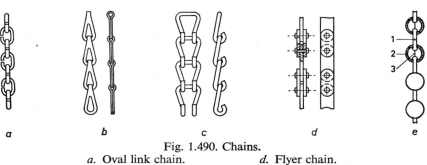

Fig. 1.490. Chains.
a. Oval link chain.   d. Flyer chain.
b. Patented chain.   e. Bead chain.
c. Hook-link (jack) chain.

1 = rod.   2 = punched and pressed bead.   3 = seam.

To provide the necessary flexibility, metal cables are built-up of several strands, each containing two or three individual wires. The materials used are steel, nickel or phosphor bronze.

Metal bands and chains are suitable for transmitting heavier loads without stretching unduly. Some of the bands are made endless and perforated to run on toothed rollers.

Fig. 1.490 shows common types of chain, namely:

● Oval link chain, with open or closed links, used in modern clocks: flexible and easily guided on rollers.
● Patented (chandelier) chain, used as a suspension or pull chain.
● Jack (or hook-link) chain: its load capacity is small: it can run on toothed rollers.
● Flyer chain: very narrow and able to withstand relatively heavy tensile loads. Used in combination with a conical drum and driving spring to supply a constant driving force (Fig. 1.491).
● Bead chain, consisting of punched beads linked flexibly by short rods: used in pull switches.

A drum is used to wind up (or take up) a driving belt, as shown in Fig. 1.492. Figure 1.493 shows a composite drum for a driving cord. Whilst this particular drum is smooth, the periphery can also be helically grooved as a guide, to prevent the turns from overlapping. The same result can also be obtained with a smooth drum by imparting a suitable traverse to this during winding. See also Section 1.11.6 (b) and Fig. 1.520.

When flexible bands are used to *transmit compressive forces* in very much the same way as tensile forces, it is necessary to ensure that the force cannot cause the band to buckle in the guide. Such guides take the form of sheaths or conduits, in which the band is adequately supported and retained (Fig. 1.494).

A variant of this is the pressure drive with balls. Figure 1.495 shows a vice in which the compressive force is transmitted via balls which can be thrust along the conduit.

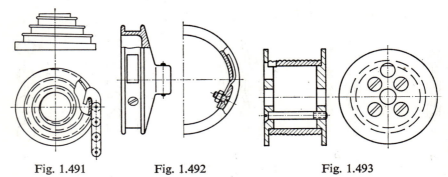

Fig. 1.491   Fig. 1.492   Fig. 1.493

Fig. 1.491. Conical drum with flyer chain and spring, to deliver a constant driving force.
Fig. 1.492. Drum to take up driving belt or band.
Fig. 1.493. Smooth composite drum for cord.

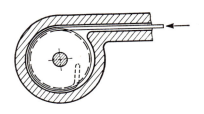

Fig. 1.494. Transmission of compressive force by sliding band.

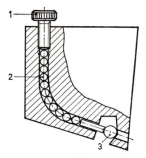

Fig. 1.495. Transmission of compressive force via balls in a suitable conduit, used on a vice.
1 = pressure screw.
2 = balls.
3 = clamped workpiece.

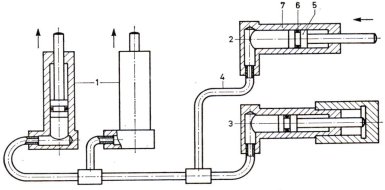

Fig. 1.496. Transmission by hydraulics.
1 = cylinders. 2 = pump. 3 = corrector. 4 = duct. 5 = piston.
6 = O-ring. 7 = cylinder.

Compressive forces can be transmitted also by driving a cable through a flexible conduit (Bowden cable). For particulars, see Section 1.4.2 and Figs. 1.251 and 1.252.

As a further development on these lines, it is possible to employ hydraulics, a very simple version of which is shown in Fig. 1.496. Here, two cylinders (1) are driven by a piston pump, and the system regulated by a piston corrector (3).

Such systems are constructed from a kit of components (cylinders, slides, valves, non-return valves and other circuit elements) to serve as controls and/or boosters in machines and appliances.

The high working pressure (about 10 N/mm²) enables substantial forces to be transmitted without bulky parts. Given flexible conduits, the work cylinder does not have to be in a fixed position relative to the pump: the two can be moved relative to one another. An adjustable safety valve protects the system against overload.

Pneumatic systems are only used for such purposes when accurate indexing between the end positions is not required.

REFERENCES

[1] RICHTER, V. VOSS and KOZER, *Bauelemente der Feinmechanik*, Verlag Technik, Berlin, 1954.
[2] K. HAIN, *Die Feinwerktechnik*, Fachbuchverlag Dr. Pfanneberg, Giessen, 1953.
[3] G. SCHLEE, *Feinmechanische Bauteile*, Verlag Konrad Wittwer, Stuttgart, 1950.
[4] D. C. GREENWOOD, *Product Engineering Design Manual*, McGraw-Hill, New York/Toronto/London, 1959.
[5] *Mechanical Drives Reference Issue*, Machine Design, 1967, September 21.
[6] P. PIETSCH, *Kettentriebe*, Konstruktion, 1954, 2, p. 46.
[7] R. W. CARSON, *Instrument Drives*, Product Engineering, 1961, October 30, p. 50.
[8] Catalogue MEBAG Micro Hydrauliek, Vlijmen.
[9] F. J. C. RADEMACHER, *Mechanische aandrijvingen*, Polytechnisch Tijdschrift W., 1968, p. 609, p. 637, p. 680, p. 743, p. 776, p. 900, p. 942.

1.11 Friction drives[1 - 12]

1.11.1 Introduction

Drives in appliances, and sometimes in machines, often take the form of friction transmission systems, transmitting force and motion through the friction set up by a driving member rolling on a driven member. Sufficient friction can only be obtained if there is enough contact force between the driving and driven members, and if friction materials having a suitably high coefficient of friction are employed.

This is the essential difference between friction transmission systems and, say, transmission by gears. Beyond a certain limit, slip is to be expected, causing the driven member to lag the driving member. The lower the coefficient of friction, the greater the normal pressure required to transmit a given torque.

Formulated:

$$M = W.R = \mu N.R,$$

where M is the torque to be transmitted,
$\quad W$ is the frictional force to be transmitted,
$\quad \mu$ is the coefficient of friction,
$\quad N$ is the normal force between the members in contact, and
$\quad R$ is the radius of the friction disc up to the point of contact.

For further details of design theory see Volume 1, Chapter 2.6.

In the absence of slip, the transmission ratio of friction wheels, like that of gears, is inversely proportional to the ratio of their diameters. By slipping, a friction drive may act as a torque limiter, and give protection against torque overload (see also Section 1.9.8).

The coefficient of friction (μ) between two interacting friction materials is not a constant. It depends on such factors as: surface condition, lubrication, relative velocity, specific pressure, humidity and temperature of the friction surfaces, and μ can vary from twice, to half, its average value.

To ensure effective transmission of the friction as planned, the normal force should be two or three times that calculated from the data concerning torque and friction coefficient.

Apart from the coefficient of friction, another factor governing the choice of materials is the noise made by the mating rollers. Cast iron is quieter than mild steel; synthetic resin bonded fabric or paper, and nylon, are even quieter. Plastics may be permanently deformed when subjected to a force continuously for some time with the drive at a standstill. The coefficient of friction depends on the relative velocity (v) of the surfaces rubbing together. To put this another way: the value of the coefficient of friction depends partly on the amount of slip (Fig. 1.497).

For most of the different combinations of materials, μ reaches a peak at a velocity $v = 0$, that is, with static friction (μ_0), falls off rapidly in response to a very slight increase in velocity, and then gradually increases with further increasing velocity, until the frictional energy generated at a very high velocity finally destroys the surface of the material. An exception to this progression is the combination of dressed leather on metal, which leads to a constant μ as regards differences in velocity between the members rubbing ogether. Leather therefore does not have any tendency to snatch (stick-slip).

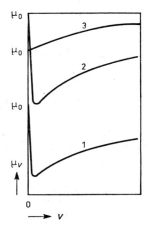

Fig. 1.497. Variation of coefficient of friction with increasing slip velocity, for different combinations of materials.
1. Metal on metal, lubricated.
2. Driving cord on metal.
3. Leather on metal.

Because the coefficient of friction increases with the difference in velocity between friction surfaces, insufficient normal force leads to a state of balance at a relatively higher slip velocity, and therefore at a higher coefficient of friction. In given circumstances, the slip velocity may therefore depend on the frictional force that can be applied and on the normal force available.

Table 1.2 lists some representative values of friction coefficients for combinations of materials commonly used in friction drives. For a reliable friction drive, it is advisable to allow a wide margin in the calculation. The coefficient of friction can be increased by knurling one of the friction members slightly.

TABLE 1.2
Representative values of friction coefficients for different combinations of materials (not lubricated)

Driving member	Driven member	μ
Leather, ground free from grease	Steel, ground	0·4
Leather	Steel, polished	0·3
Rubber	Steel, ground	0·45
Rubber	Steel, polished	0·3
Rubber	Paper	0·4
Steel, polished	Paper	0·2
Cast iron, polished	Rubber	0·3
Brass, knurled	Paper	0·4
Steel	Vulcanized rubber	0·5
Brass	Zinc, knurled	0·5
Fibre	Steel, ground	0·4
Synthetic resin bonded paper Synthetic resin bonded fabric	Steel, ground	0·25

Permanent deformation and wear must be taken into account. Fibre does not reveal much change in coefficient of friction after prolonged use, but is unsuitable for some purposes, because it is hygroscopic. The coefficient of friction of cast iron used with synthetic resin bonded paper or fabric shows little variation.

1.11.2 Friction drives with parallel shafts

Friction wheels are preferred for precision adjustments of, say, variable capacitors, as a simple means of obtaining the exact tuning required with a high transmission ratio and a minimum of backlash.

The shaft of steel or brass in Fig. 1.498 is finely knurled to give it a good grip ($\mu=0·5$) on the driven zinc or aluminium disc. The normal force is sustained by a tension spring. For greater friction, the shaft can be grooved instead of knurled, to increase the frictional force by increasing the effective normal pressure. The increase thus obtained is a factor of $1/\sin \delta$, where δ, the half-apex angle, is between 10° and 15° in Fig. 1.498b. The angle of the disc has been taken larger than the angle of the groove, to ensure general two-point contact between shaft and disc and thus avoid undue wear owing to slip. The drawbacks of this construction are that:

● The vernier shaft is not fixed accurately in place, so that the drive may be interrupted by clumsy handling.
● If the disc is driven by direct manipulation for coarse tuning, fine tuning and coarse tuning involve rotation in opposite directions.
● There is likely to be more wear with the shaft seated in an elongated hole than in a round hole.

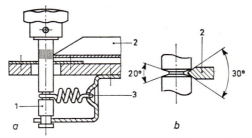

Fig. 1.498. Friction drive for precision adjustment
a. Friction increased by knurl on shaft.
b. Friction increased by groove in shaft.
1 = driving shaft.   2 = driven disc.
3 = tension spring.

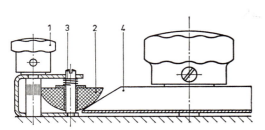

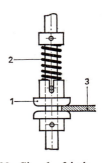

Fig. 1.499. Friction drive with rubber idler.
1 = driving shaft.   3 = helical spring.
2 = idler.   4 = driven disc.

Fig. 1.500. Simple friction
wheel drive.
1 = driving roller.
2 = helical spring.
3 = driven wheel.

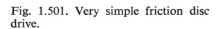

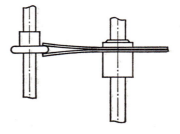

Fig. 1.501. Very simple friction disc
drive.

Figure 1.499 shows a drive without these faults. The rubber idler is preloaded against the knurled shaft, and is held against the disc by a helical spring. A simple friction drive appears in Fig. 1.500, where the normal force is supplied by a helical spring fitted on the driving shaft, which clamps the wheel between two driving rollers of hardened steel.

In the still simpler construction of Fig. 1.501, the normal force is derived from the resilience or elasticity of two relatively large friction discs which, before assembly, are flat, with a slight radius at the edges.

These discs are made of spring steel strip, to avoid deformation after prolonged use. The small friction wheel is hardened and polished to avoid erosion and undue wear.

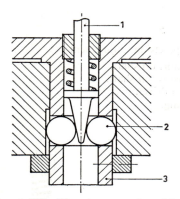

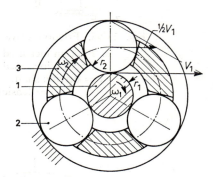

Fig. 1.502. Planetary vernier drive.
1 = driving shaft.
2 = three balls.
3 = driven shaft.

Fig. 1.503. Determining the transmission ratio of a planetary vernier drive.
1 = driving shaft.
2 = three balls.
3 = driven shaft.

1.11.3 Friction drives with concentric shafts

Concentricity of the driving and driven shafts is exemplified by the planetary ball vernier drive of Fig. 1.502. The balls, usually three, function as planet wheels and are clamped between the driving shaft and the stationary housing by a spring. When the shaft rotates, the balls are driven by friction and roll against the stationary housing. The ball cage, connected to the driven shaft, is carried along. To stop the transmission of motion, the shaft must be retracted against the thrust of the spring.

Fig. 1.503 illustrates the method of determining the transmission ratio. With the driving shaft rotating at an angular velocity ω_1 and with a peripheral velocity v_1 at the position of the balls, the peripheral velocity of the ball centres is $\frac{1}{2}v_1$, associated with an angular velocity of the driven shaft ω_2, and measured at a radius $r_1 + r_2$. Thus,

$$\tfrac{1}{2}v_1 = \omega_2(r_1 + r_2) = \tfrac{1}{2}\omega_1 r_1$$

and the transmission ratio is

$$\frac{\omega_1}{\omega_2} = \frac{2(r_1 + r_2)}{r_1}.$$

In the extreme case that $r_2 = 0$, the transmission ratio is a minimum, and equals 2. Usual transmission ratios are between 2·5 and 6.

A drawback of the construction in Fig. 1.502 is that the driving shaft must not be acted on by an axial force that would oppose the spring pressure and thus affect the torque transmission adversely. Another drawback: it is often difficult to position the centre-lines of the ball seats (holes) in a plane at right-angles to the centre-line of the shaft, as must be done to ensure smooth and torsionless running of the vernier drive. The points of frictional contact between the balls and shaft should therefore be located on the same peripheral circle of the cone. Fig. 1.504 shows a system not subject to these drawbacks. Here, no axial force is exerted on the ball cage to balance the spring pressure, so the shaft cannot affect the ball friction. See also Volume 7, Section 2.5.2.

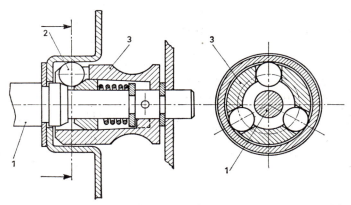

Fig. 1.504. Vernier drive in which the shaft does not affect the ball friction.

1 = driving shaft.   3 = driven shaft in the form of
2 = three balls.   slip roller for cord.

Advantages of planetary ball vernier drives compared with other precision controls are:
- They can be accommodated in a small space, or built into a bearing if necessary.
- They need fewer bearing points, due to the absence of a parallel shaft.
- They can be fitted on the shaft stub of a driven member.
- There is no reversal of direction of rotation.

The three-stage variator, or variable-ratio transmission, shown in Fig. 1.505, is a friction drive with a transmission ratio continuously variable over a very wide range. In it, several coned discs are employed, which can rotate independently on a common shaft, with the exception of the extreme left-hand (1) and right-hand (7) discs, which are keyed to the driving and driven shafts, respectively. Every second disc (4) carries three balls at angles of 120°, which can be thrust outwards and towards the centre by means of the lever (10), used to adjust the transmission ratio. When disc (1) rotates the first group of three balls, disc (3) is driven at a slower speed due to the difference in the radii r_1 and r_2. Disc (3) then drives disc (5), which in turn drives disc (7), each with the same speed ratio. The final speed of the driven shaft (9) depends on the product of the three transmission ratios. Thus, a compact variator with a wide range of speed variation is obtained.

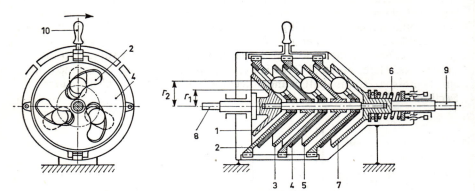

Fig. 1.505. Three-stage frictional drive with continuously variable transmission ratio giving a wide range of speed control.

1 = disc keyed to driving shaft (8).
2 = disc to hold three balls at angles of 120.°
3 = disc free to rotate on shaft.
4 = disc on speed control lever (10).
5 = disc free to rotate on shaft.

6 = compression spring.
7 = disc keyed to driven shaft (9).
8 = driving shaft.
9 = driven shaft.
10 = speed control lever.

1.11.4 Friction drives with intersecting shafts

Fig. 1.506 is an example of such a transmission with a fixed transmission ratio and with the shafts intersecting at right-angles. This conical wheel drive is used to operate film take-up spools in a projector. The driving pinion is made of synthetic resin bonded fabric, or of leather, and is drawn against the cast iron driven wheel by a spring. This transmission is quiet, readily engaged and disengaged, is free from backlash, and can be used as a slip coupling or clutch. Drawbacks are the unfavourable bearing load and the spring power required. In practice, it is difficult to comply with the requirement governing pure rolling motion, namely, that the line of contact should pass through the point of intersection of the two shafts. For this reason, the two wheels are often only brought into contact along a section of the contact line. Fig. 1.507a gives an example of this. The transmission continues to function correctly even when shaft (1) does not pass exactly through the apex of the cone wheel. Hence, right-angle friction drives, with a flat disc as the driven wheel, are employed (Fig. 1.507b).

It is a simple matter to design friction drives in such a way that the transmission ratio between two shafts is continuously variable, or the direction of rotation of the driven shaft is reversible. Fig. 1.508 shows several such examples, involving an additional, movable friction roller.

In Fig. 1.509*a* and *b*, the friction roller is revolved. Fig. 1.510 shows how the direction of rotation is reversed by shifting a hub with two discs. Such reversal can be accomplished too by sliding the driving friction wheel beyond the centre of the driven disc.

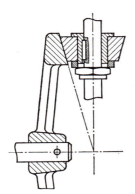

Fig. 1.506. Friction transmission with conical wheels.

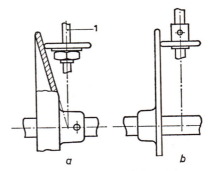

Fig. 1.507. Simple friction drive.
a. With conical wheel and (disc shaped) friction roller.
b. With flat disc and friction roller.
1 = driving shaft.

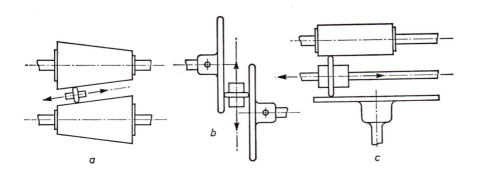

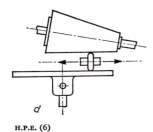

Fig. 1.508. Friction drive with continuously variable transmission ratio, obtained with sliding friction roller.

  a. Between two conical wheels.
  b. Between two flat discs.
  c. Between cylindrical wheel and flat disc.
  d. Between conical wheel and flat disc.

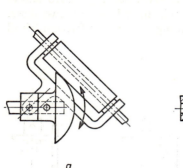
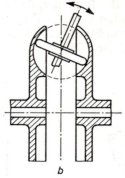
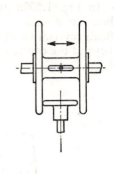

a   b

Fig. 1.509. Friction drive with continuously variable transmission ratio, obtained with pivoted friction roller.

    a. In combination with spherical wheel.
    b. Between two discs with toroidal recess.

Fig. 1.510. Direction of rotation reversed by means of friction roller and two sliding discs.

1.11.5 *Friction transport and feed mechanisms*

One example of a friction transport mechanism is a miniature locomotive running on rails. An indoor transport-system for letters is similar in principle (Fig. 1.511). The frame supports the bearings of the traction wheels, the largest of which is driven, thereby moving the trolley along the two taut wires. The guides on each side of the lower wire prevent tipping.

Friction also plays a part in feed mechanisms, e.g. the paper feed of a typewriter (Fig. 1.512).

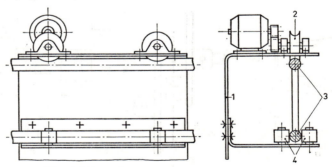

Fig. 1.511. Indoor transport system for letters.

1 = frame.   3 = two taut wires.
2 = driving traction wheel.   4 = guide.

In Fig. 1.513, the axially-retained roller drives a skew wheel which is spring-loaded against the roller and shifted sideways by it. At the end of the stroke, the wheel swivels so as to transverse in the opposite direction.

The mechanism in Fig. 1.514 is a more refined version, involving special ball bearings whose inner shaft ring is arched clear of the shaft. The driving shaft is encircled by three of these bearings, in such a way that the middle bearing is pressed down on the shaft whilst the other two cushion the rebound, so that the mechanism in itself is load-relieved. Given suitable contact between shaft and rings, fairly heavy loads can be transmitted by a small mechanism. The skew setting of the bearings results in traversing motion when the driving shaft rotates. At the end of the stroke the bearings tilt over to reverse the traverse.

In Fig. 1.515 a revolving shaft sets a sleeve in sliding motion by means of balls seated in a number of separate grooves. These balls are clamped between shaft and sleeve, except through a certain angle α, within which the gap between shaft and sleeve is wider. Within this angle, the balls are returned before moving forward again in the rest of the groove and taking the sleeve with them. Here, the pitch of the movement is not adjustable, as it is in the other two mechanisms.

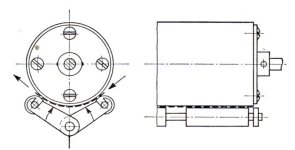

Fig. 1.512. Paper feed mechanism in typewriter.

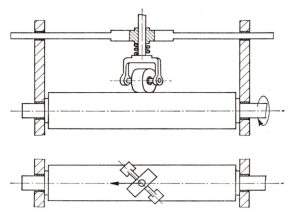

Fig. 1.513. Friction transport mechanism with skew friction roller driven by cylinder, to supply reciprocating motion.

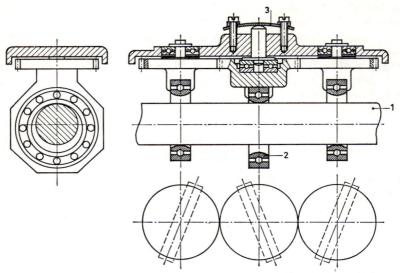

Fig. 1.514. Friction transport mechanism with three skew ball bearings whose arched inner ring is driven by shaft to supply reciprocating motion.

1 = driving shaft.
2 = ball bearings with arched inner ring.
3 = loading spring.

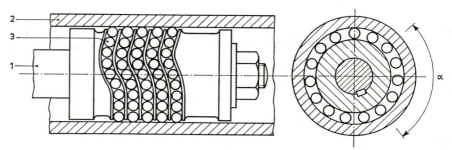

Fig. 1.515. Sliding motion of sleeve obtained by means of revolving shaft with balls in separate, specially-shaped grooves.

1 = driving shaft.   2 = sleeve.   3 = balls.

1.11.6 Cord and belt transmission systems

(a) Introduction

This section deals with transmission systems in which engagement of the driven member is not positive, as it was in Section 1.10, but frictional. In precision mechanisms, the following are used for transmitting the tensile load; hemp, cotton or silk cord and nylon rope or catgut 0·6 mm or more thick, depending on the forces involved. For greater forces, e.g. in small machines, round or flat laminated plastics belts are used (see Figs. 1.516 and 1.517). The belt should not stretch appreciably, but should not lack flexibility and be very hard-wearing, with a fairly high coefficient of friction relative to the material used for the friction driving rollers, which is generally brass or nickel-plated steel.

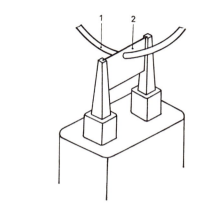

Fig. 1.516. Welding the ends of a round plastics belt together by means of a heated blade.
1 = round belt.   2 = heated blade.

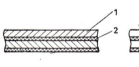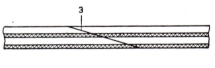

a   *b*   *c*

Fig. 1.517. Flat, laminated plastics belt.

a. Cross-section.   1 = layer of plastic.   3 = bonded adhesive
b. Shape of weld.   2 = layer of fabric.   joint
c. Honeycombed surface.

Vee-grooved sheaves are used to guide ropes and cables. The groove is shaped to minimize squeezing and deformation of the cable or rope, with an included vee-angle from 60° to 90° and a groove radius of about half the cable thickness. Guiding sheaves for metal wires or cables, often constituting precisely that connecting span of the tensile linkage in which length variations can least be tolerated, are given a larger diameter than those used for cord. This is necessary to limit the bending stress of the metal cable, with its high modulus of elasticity. For the same reason, the cable is made up of several strands, each containing several individual wires.

Where a rope or cable sheave has to transmit an effective torque, a large area of contact is desirable and the included angle is 60° (Fig. 1.518*a*). A sheave of the kind shown in Fig. 1.518*b*, with an included angle of about 45° and a small radius, grips well, and provides more torque from the motion of the cable.

Sheaves used in sets and appliances are usually made of moulded thermo-setting plastics, extruded thermoplastics or turned metal (e.g. brass) and are sometimes assembled from blanked metal parts (Fig. 1.518c).

Factors giving easy running of a sheave are:

- High ratio of sheave diameter to sheave trunnion diameter.
- Suitable combination of sheave and trunnion materials.
- Suitable surface conditions, specific pressure and lubrication of the bearing surfaces.
- Adequate sheave diameter, having regard to the stiffness of the cable.

In cheap drives, or where space is limited, smooth, stationary pins are sometimes used instead of sheaves and cause a great deal more friction. Pin diameters range from 5 to 8 mm for use with 1-mm thick cord. Aluminium is unsuitable for cord guides.

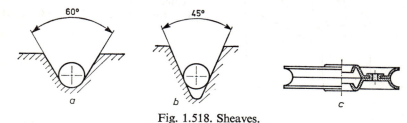

Fig. 1.518. Sheaves.

a. Groove profile giving large area of contact.   *c.* Composed of blanked parts.
b. Groove profile in which cord sticks.

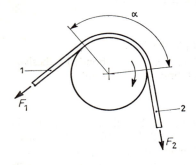

Fig. 1.519. Determination of the ratio of the two side forces of a cord slipping round a shaft or roller.

F_1 = force on side 1.
F_2 = force on side 2.
α = encompassed angle.

(b) *Shafts for friction drive with cord*

They are grooved or fitted with a special slip roller on which the cord runs. From such a shaft, both sides of the cord run to the driven member and, if necessary, to an indicator also.

Theoretically, the maximum ratio (f) of the two side forces on a revolving cylindrical shaft (Fig. 1.519) is:

$$f = \frac{F_1}{F_2} = e^{\mu\alpha}$$

where F_1 is the force on side 1;

F_2 is the force on side 2;

e is the base of the natural logarithm;

μ is the coefficient of friction between cord and roller;

α is the angle encompassed by the cord (radians).

This holds for the direction of rotation indicated in the diagram. In practice the side force ratio (f) is smaller than the formula suggests, mainly owing to the stiffness of the cord, which affects the normal force adversely.

Another uncertain factor in the formula is the coefficient of friction μ, which depends not only on the different materials and the surface condition of cord and shaft (or roller), but also on the slip velocity of the cord. With the drive running, a certain amount of slip is inevitable, because the cord stretches in response to increased tension as it passes from side 1 to side 2. This rules out static friction μ_0 (Fig. 1.497), but leaves μ_v, increasing with the slip velocity, which may lead to a state of balance between the transmissible tangential force on the shaft or roller and the frictional force increasing with the slip velocity.

Plain brass or nickel-plated steel are the materials most used for the driving shaft or slip roller, the surface roughness of which should not exceed 15 ru (microinch) if undue cord wear is to be avoided. Experience has shown that a used slip surface on a shaft, roller or cord acquires a very much lower coefficient of friction, by becoming smoother as a result of slip.'

To obtain good driving torque, the cord is wrapped two, three, or at most, four times round the shaft or roller. With the drive running, the cord moves in the axial direction and takes up space in the structure. This can be limited by causing the cord to ride-up against a collar on the shaft or slip roller in such a way as to slide back regularly without catching, that is, without the turns overlapping, as determined by the radius of the collar (Fig. 1.520). The tangential frictional force set up by slipping as the cord slides back is smaller than when the turns are on the cylindrical portion of the shaft or roller. From this point of view, the roller in Fig. 1.520a is correctly shaped, since it limits the need for slide-back. Fig. 1.520c gives the best assurance against snagging.

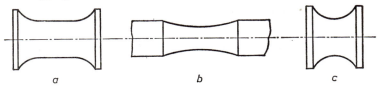

a b c

Fig. 1.520. Slip roller profiles, with collar to limit the axial motion of the cord turns.
    a. With cylindrical part which limits the need for slide-back of the turns.
    b. Turns slide back regularly: prone to snagging.
    c. Turns slide back regularly: snagging is avoided.

(c) *Belts* (see also Section 1.10)

Round belts fall into two main groups:
- Those woven out of synthetic yarn (polyamide).
- Those made of solid plastics (polyurethane rubber or polyvinyl chloride) with diameters from 2 to 15 mm.

The stranded (built-up of twisted strands) round belt is endless and seamless and must be ordered to exactly the length required. The solid belt is readily welded and stored in long lengths on spools. Welding is done by pressing the two ends to be joined against a heated blade and then withdrawing the blade (Fig. 1.516). These belts are flexible in all directions without stretching and possess sufficient adhesion when used with pulleys made of the customary materials. They are employed in light engineering projects involving flexure in several directions, e.g. engraving machines, textile machinery and electrical engineering equipment.

For all practical purposes, leather belts are no longer used in precision mechanisms and machines; they have been superseded by belts built up of one, two or three layers of plastics laminated with fabric woven from synthetic yarns or fibres (for instance, a combination of chloroprene and polyurethane, reinforced with polyamide). These belts, up to about 5 mm thick (Fig. 1.517) are strong and flexible and have a roughened (honeycombed), hard-wearing surface that clings well to a pulley. A slightly convex (motor) pulley is excellent for holding the belt in position. Suitable equipment for the preliminary treatment and finishing of adhesive joints is available. Such belts can be run at fast speeds and are used to drive grinding heads, diamond drilling machines (Volume 3, Chapter 4) and finger-tip drilling machines (Volume 10, Chapter 2).

Flat belts, or belts round or square in section, made of chloroprene rubber (Neoprene) or polyurethane rubber (Vulkollan and Genthane), are used in cheap, lightweight drives for, say, playback equipment (tape recorders). These endless belts are cropped from a hat-shaped injection moulded piece and in many cases finish-ground to remove burrs and eliminate any troublesome deviations of form.

For particulars of vee-belts and other belts for heavier duty in engineering, see the appropriate handbooks.

REFERENCES

[1] RICHTER, V. VOSS and KOZER, *Bauelemente der Feinmechanik*, Verlag Technik, Berlin, 1954.
[2] K. HAIN, *Die Feinwerktechnik*, Fachbuch Verlag Dr. Pfanneberg, Giessen, 1953.
[3] G. SCHLEE, *Feinmechanische Bauteile*, Verlag Konrad Wittwer, Stuttgart, 1950.
[4] D. C. GREENWOOD, *Product Engineering Design Manual*, McGraw-Hill, New York/Toronto/London, 1959.
[5] K. RABE, *Grundlagen Feinmechanischer Konstruktionen*, Anton Ziemsen Verlag, Wittenberg/Lutherstadt, 1942.
[6] R. W. CARSON, *Instrument Drives*, Product Engineering, 1961, October 30, p. 50.
[7] *Mechanical Drives Reference Issue*, Machine Design, 1967, September 21.
[8] W. PEPPLER, *Zweiachsige Reibradantriebe für feste Übersetzungen*, Konstruktion, 1949, p. 289 and 336.
[9] G. Niemann, *Reibradgetriebe*, Konstruktion, 1953, 2, p. 33.
[10] E. CH. KRAUS, *New Approaches to Variable Speed Drives*, Machine Design, 1953, December, p. 232.
[11] A. GEMANT, *Frictional Phenomena*, Chemical Publishing Co., U.S.A., 1950.
[12] S. HILDEBRAND, *Feinmechanische Bauelemente*, 1968, Carl Hanser Verlag, Munich.

1.12 Gear wheels and gear transmissions [toothed][1 – 6, 14, 34]

1.12.1 Introduction[7 – 9, 12, 13, 16, 21, 24, 31, 32]

Unlike friction wheels, gear wheels in transmission systems are coupled through the meshing of mating shapes. This rules out slipping altogether. In principle, gear wheels and gear teeth are governed by the laws of conventional engineering, but the requirements for precision engineering shift the emphasis somewhat. For instance, the main concern is with transmission systems that are slow running; have to transmit only minor forces; and are permitted some variation in the transmission ratio as they pass a tooth.

For this reason, they are usually cheap, as indeed they have to be. Again, gears with only slight backlash, or sometimes free from backlash, are needed in a number of cases. This can be accomplished in various ways. For example, when a precision transmission system with a perfectly constant angular velocity ratio is required for measuring purposes, the tooth flank should be as close as possible to the theoretical profile, since this is designed on the basis of a constant speed ratio between interacting members.

(a) *Involute teeth*

In precision engineering, as in conventional engineering, gears are almost invariably involute, because:

- The simple rack shape simplifies the basic tooling.
- Accurate machining is possible.
- Precise engagement is ensured, despite minor changes in spacing.
- Teeth of this kind are suitable for change wheels.

A pressure-angle of 20° has been adopted for most purposes. The pressure-angles of 14° 30′ and 15° formerly employed have been abandoned because, towards fewer teeth, a small pressure-angle imposes a higher limit on the minimum number of teeth obtainable without undercutting. In choosing the 20° pressure-angle, it has been necessary to forgo the advantages of slightly higher efficiency, quieter running and a higher contact ratio.

For further particulars on involute gear teeth, and the cycloidal gear teeth mentioned later, see Volume 1, Section 2.5. For details of standard basic rack profiles and modules, see Volume 1, Section 1.9.

An important aspect of gear design is that the gear often has to be accommodated in a small space. If helical gearing cannot be used, this leads to small modules (to 0·1 mm), possible only in low-torque transmissions. When, as often happens, the driving torque is only slightly greater than the product of the resistive torque and the transmission ratio, it is most important for the gearing to run lightly despite dirt, metal dust caused by wear, or resinous residues from the lubricant. In the case of involute gear teeth, the remedy is to increase the bottom clearance and the flank clearance, which also makes the process of gear cutting less critical. To avoid detracting from the contact ratio, the addendum is also increased, to 1·2 module (m) (NEN 1628 and DIN 58400). In this way, the centre distance is increased by the equivalent of twice the increase in the addendum (Fig. 1.521).

Gears having a high transmission ratio (up to about 1 : 10) are used more in precision, than in conventional, engineering. Here, everything depends on the ability to reduce the number of teeth on the pinion to a minimum. From this point of view, involute gearing is at a disadvantage. With a pressure-angle of 20° and an addendum equal to the module, at least 16 teeth are required to mesh with a wheel having 111 teeth or fewer, without undercutting. Position correction of the pinion enables the minimum number of teeth to be reduced considerably, namely to 7, but not to the extent possible with cycloidal gearing, where there is no difficulty whatever in using 6 teeth without undercutting and where the line of action (or path of contact) is favourable enough to ensure a sufficiently high contact ratio. This, together with the relative insensitivity to dirt, explains the preference for cycloidal teeth in clocks, although the number of teeth on involute helical gears can be similarly reduced by employing a small addendum (0·8 m). The reason for not using helical gearing is that manufacturers are not prepared to accept the drawbacks, namely, that such gears are more difficult to cut and that the tooth pressure has an axial component (see also Section 1.12.3 (a)).

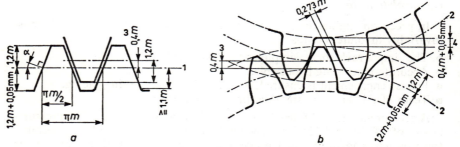

Fig. 1.521. Involute teeth with enlarged bottom clearance and increased addendum (deep profile) for modules from 1 up to 20.
    a. Theoretical rack profile with rack profile of the counter wheel.
    m = module; πm = pitch; α = pressure angle = 20°.
    b. Two wheels rolling in contact.
    1 = base line; 2 = pitch circle; 3 = extra space between normal pitch circles; 4 = extra large bottom clearance.

(b) *Cycloid teeth*

In watches the diameter of the generating circle of cycloid gearing equals half the pitch circle of the gear wheel or pinion. The root of the tooth is thus limited by straight, radial flanks (Fig. 1.522). The teeth of the wheel are designed heavier than those of the pinion:
- tooth thickness: pinion 1·14 m; wheel 1·57 m;
- width of tooth space: pinion 1·14 m; wheel 1·57 m;
- flank clearance at circumference 0·43 m.

The material used for the wheel is brass and that for the pinion is steel, so the strength is also taken into account. The wheels use the full available part of the epicycloid as flank for the addendum, thus the tip is sharp. The addendum varies between $1 \cdot 4\,m$ and $1 \cdot 8\,m$ and increases with the transmission ratio and the number of teeth of the pinion.

This broad, sharp and thus high tip ensures that when two teeth move out of mesh the next pair of teeth is already near the centre-line between the centres of the two cycloidal wheels (Fig. 1.523). Thus the unfavourable situation of Fig. 1.524 is avoided. As used in watches, the wheel drives the pinion and the tooth (4) of the pinion encounters the frictional force exerted on the movement. This hardly applies for Fig. 1.523.

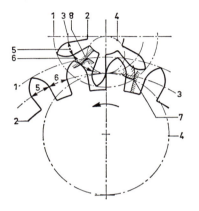

Fig. 1.522. Cycloidal teeth with ample flank clearance.

    1 = tip circle.
    2 = root circle.
    3 = pitch circle.
    4 = rolling circle.
    5 = tooth thickness.
    6 = width of tooth space.
    7 = start of meshing.
    8 = end of meshing.

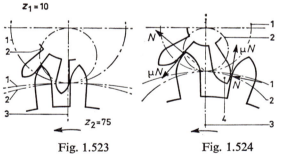

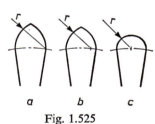

| Fig. 1.523 | Fig. 1.524 | Fig. 1.525 |

Fig. 1.523. When two teeth move out of mesh, the next pair of teeth is already near the centre-line between the centres of the two cycloidal wheels.
    1 = pitch circle.   2 = rolling circle.   3 = centre line.
Fig. 1.524. Pinion encounters resistive torque because of the frictional force μN exerted by the driving cycloidal gear wheel on tooth (4).
    1 = pitch circle.   3 = centre line.
    2 = rolling circle.   4 = tooth of pinion.
Fig. 1.525. Usual, simplified tooth tips on cycloidal pinions.
    a. Pointed tooth: r = tooth thickness.
    b. Semi-pointed tooth: r = 5/6 tooth thickness.
    c. Round tooth: r = 1/2 tooth thickness.

The high contact ratio enables part of the line of action to be dispensed with. Omitting the sector where the tip of the pinion makes contact enables the shape of this tip to be simplified (Fig. 1.525). This tip can be bounded by one or two arcs of a circle which mate exactly with the epicycloids at the pitch circle but encroach inside them elsewhere. Thus, the contact ratio is invariably slightly greater than unity. However, the pointed tooth is retained in very accurate movements (Fig. 1.525a), whereas the semi-pointed (Fig. 1.525b) and round tips (Fig. 1.525c) are used in ordinary clocks. Of the three simplified shapes, the round one gives the greatest saving in terms of material and tools. The principal requirement is that the torque transmitted should vary as little as possible.

Drawbacks of cycloidal gear teeth are:
- The centre distance cannot be changed.
- A wider range of more complex tools is required.
- Accurate machining is difficult because the flank profile is made up of two different curves.
- The shape of the tooth is not suitable for blanking, since the tip is too pointed and the root too narrow.
- They are not suitable for change wheels, because this calls for rolling circles of equal size.

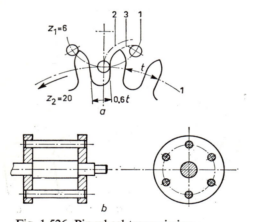

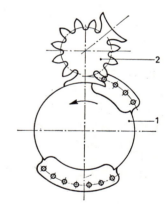

Fig. 1.526. Pin wheel transmission.
a. Pin wheel and gear wheel in mesh.
b. Construction of pin wheel.
1 = pitch circle.
2 = epicycloid.
3 = equidistant of the epicycloid.
t = pitch.

Fig. 1.527. Star wheel drive.
1 = pin wheel.   2 = star wheel.

As we have seen, cycloidal teeth give considerable scope for variation. One version worth mentioning is the pin wheel transmission (Fig. 1.526), a simple and very old-established mechanism found in clocks, toys and in the engineering industry. It runs efficiently, even in accelerated drives, and requires remarkably little attention even in dusty surroundings. Friction can be reduced by using rollers instead of fixed pins, but this is conditional on good lubrication. The tooth flanks on the wheel are equidistant from the epicycloids described by the pin centres. The tooth space is semicircular, with a diameter of $0 \cdot 6\ t$ (pitch). The line of action is unilateral. The tooth crest is raised high enough to ensure a contact ratio greater than unity.

Teeth of this kind are used in star wheel drives too. In Fig. 1.527 the pin wheel drives the star wheel and then leaves this motionless for a time. The radii of the pitch circles of the two groups of pins are different, in order to satisfy the requirements imposed.

(c) *Choice of material*

In precision engineering, the choice of materials is governed mainly by:
- The method of manufacture and the associated costs.
- The resistance to wear.
- The resistance to corrosion.

Because the dimensions, and therefore the amount of material used, are generally small, material costs are relatively unimportant.

As in ordinary mechanical engineering, unlike materials are chosen for mating gears, the smaller wheel being the more hard wearing of the two. A much-used combination is steel and brass, with steel for the smaller wheel. Steel is used for pinions and worms and for those wheels that must combine small dimensions with great strength and limited wear. Accordingly, these parts are often hardened as well. Brass is used mainly for slow-running gears and to transmit small forces, as in clocks, musical boxes, counting mechanisms and so on. Bronze is employed instead of brass where wear must be a minimum. Steel is usually chosen as the mating material. White metal, or anti-friction metal, with a high percentage of tin, diecast zinc and aluminium are preferred for transmitting small forces in the counting mechanisms of tachometers, mileage indicators, electricity supply meters, etc. Leather, fibre, synthetic resin-bonded paper and fabric are used, as in conventional mechanical engineering, for quiet drives. Plastics, such as polyamides (nylon) and polyacetal (Delrin), are being adopted more and more.

(d) *Efficiency*

Cylindrical spur (or straight toothed) gears are the most efficient. Given good workmanship, alignment, bearings and lubrication, an efficiency $\eta = 0 \cdot 90$ is assured. The efficiency depends very much on the pressure-angle and the tooth face friction. Reduction gears are more efficient than accelerated gears.

Means of reducing the loss owing to tooth face friction are:
- Many teeth, that is, teeth that are small in relation to the wheel diameter.
- In effect, this amounts to shortening the path of contact and thus eliminating the largely sliding portions.
- Internal teeth. Such gears are more difficult to cut and support, however, and are therefore relatively expensive.

- A low coefficient of friction, or in other words sound surface conditions, good lubrication and a correct combination of materials.
- Reduction of the contact ratio (a pressure-angle of 20° is better than 15°). But this makes little real difference to spur gears, because their contact ratio is already limited.

Using helical, instead of spur gears, reduces the contact ratio appreciably, but does not create any problems, because of the overlap involved. At the same time, the gain in terms of reduced frictional loss is nullified by increased pressure on the slanting tooth. The following worm efficiencies can usually be obtained: $\eta_{worm} = 0.6$ (single start), 0.65 (two-start), 0.7 (three-start) and 0.8 (four-start).

Whilst a very much higher efficiency can be obtained with two pairs of spur gears, they take up more room and make more noise.

(e) Gear noise[30]

Noise can be cured by:
- Very accurate workmanship of finish and assembly.
- Suppression of percussive and inertial forces, amongst other things by helical gearing, insertion of elastic members to transmit power from gear ring to wheel body, and flexible couplings to unite with the adjoining machine members.
- Damping, e.g. with lubricating oil or by plastics wheels or gear rings.
- Prevention of propagation and resonance, amongst other things by avoiding bell-shaped wheels, by stiffening flat walls and by interposing sound-absorbing material.
- For methods of cutting gear teeth, see Volume 10, Section 2.7.

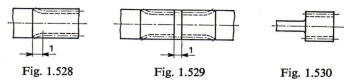

Fig. 1.528   Fig. 1.529   Fig. 1.530

Fig. 1.528. Pinion milled in shaft.
      1 = milling cutter run-out.
Fig. 1.529. Two pinions milled simultaneously in shaft.
      1 = allowance for cutting off.
Fig. 1.530. Milled pinion with root circle diameter larger than shaft diameter.

1.12.2 Spur gearing with parallel shafts[11, 15, 19]

(a) Pinions

Many different types of pinion are known. The choice is governed by considerations of space, production method, resistance to wear, strength and cost.

Machined pinions (Fig. 1.528) are milled in the shaft (of a motor, for example). This method is too expensive for longer production runs. Pinions cannot be machined more than two at a time (Fig. 1.529). Cutting-off after milling causes a horizontal burr in the tooth space which is difficult to remove. Good shaft material may be difficult to machine, so a rough face is obtained. Moreover, the milling cutter run-out may present a problem from the point of view of the bearing. If the root circle diameter is larger than the shaft diameter, the pinion can be milled according to Fig. 1.530. A drawback of this method is that often a considerable part of the shaft length has to be machined. The advantage of the versions shown in Figs. 1.528 and 1.530 is that the pinions can be made to a fine tolerance, as regards wobble of pinion relative to shaft.

Unwinding or generating several pinions at a time on a mandrel is a cheaper method.

One or two methods of fixing the pinion on the shaft are: interference fit (Fig. 1.531), plastic or adhesive joint (Loctite, Araldite) (Fig. 1.532), dowelling (Fig. 1.533), two pointed or cup-ended grub screws at 120° (Fig. 1.534), and screwing by means of extra ring on a thin-walled hub Fig. 1.535). A drawback of these methods is that the teeth may wobble relative to the bearing. This wobble may be more serious than with the pinions milled in the shaft. The least wobble is obtained by the fixing method of Fig. 1.532, if the gluing is done in an accurate jig. A point to watch when using Loctite is the removal of surplus liquid, since it does not set in the air and may therefore give trouble later by seeping into a bearing.

When pinions are made by unwinding or extruding bar material, the subsequent cutting off results, as we have seen, in burr.

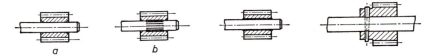

Fig. 1.531   Fig. 1.532   Fig. 1.533

Fig. 1.531. Fixing pinion on shaft by means of:
   a. Interference fit.   *b.* Knurled shaft forced in.
Fig. 1.532. Pinion fixed to shaft by adhesive joint.
Fig. 1.533. Pinion dowelled to shaft.

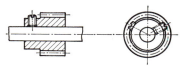

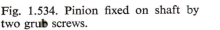

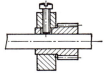

Fig. 1.534. Pinion fixed on shaft by   Fig. 1.535. Pinion fixed on shaft by
two grub screws.   two screws and extra ring.

Pinions, particularly for light loads, can be made by methods other than *machining*, such as cold working, sintering and extrusion (metal and plastics). Plastic pinions should preferably be given a land flash for dividing the mould (Fig. 1.536). The method of fixing them on the shaft may present problems.

In principle, the methods shown in Figs. 1.531 to 1.535 can be used for metal pinions. When interference fits are employed, the relative weakness of the material must be taken into account. To allow for an interference fit, or to enable, say, a gear wheel to be fastened to the pinion by rotary riveting, a metal hub can be incorporated as an insert in moulding or injection moulding. Fig. 1.537 shows one or two practical examples of such hubs.

In drives restricted to one direction of rotation, a shaft stub threaded like a wood screw, and therefore self-tapping, can be screwed in Fig 1. 538.

Because of the possibility of wobble, both pinion and stub are manufactured to very strict requirements. The shaft can also be embedded as an insert in injection moulding, which at once solves all the problems of fastening. Examples are shown in Fig. 1.539.

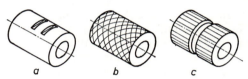

Fig. 1.536. Plastic pinion.
1 = land flash.

Fig. 1.537. Metal insert as hub for plastic pinion, anchored by:
a. Two saw cuts: a single saw cut would allow possible axial play due to shrinkage.
b. Diamond knurl.
c. Recess and straight knurls.

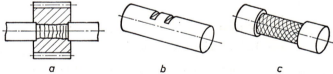

Fig. 1.538. Pinion screwed on to shaft stub threaded as a wood screw.

Fig. 1.539. Injection moulded pinion with shaft as insert, anchored by:
a. Rough-turned surface of recess in shaft.
b. Two saw cuts in shaft: a single saw cut would allow possible axial play due to shrinkage
c. Diamond knurl on recessed surface of shaft.

(b) *Gear wheels*

As a rule, they are thin in relation to their diameter. The mass of the wheel can be reduced by making holes or constrictions in the wheel body. The wheels can be turned from full-length stock and then milled on single- or multi-spindle machines, or made from round blanks milled in stacks.

Gear wheels are fixed to the shaft in the same way as pinions (Figs. 1.531 to 1.534 inclusive). Generally, when turning from full-length stock, a large amount of material has to be removed to form a hub (Fig. 1.540). It is often cheaper to fasten a separate bush in the wheel instead, by pressing, soldering, riveting or spinning over (Fig. 1.541).

Gear wheels can also be made by processes other than machining, e.g. by blanking, sintering or extruding, or by injection moulding metals or plastics.

When the wheels are blanked, the blanking arris (or detruded edge) must be taken into account (Fig. 1.542). (See also Fig. 1.66).

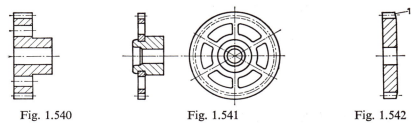

Fig. 1.540   Fig. 1.541   Fig. 1.542

Fig. 1.540. Gear wheel turned from full-length stock.
Fig. 1.541. Gear wheel fastened on hub by rotary riveting.
Fig. 1.542. Blanked gear wheel.   1 = arris.

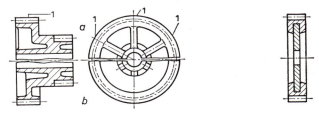

Fig. 1.543. One - piece injection moulded gear wheel and pinion.
a. Deformation caused by uneven shrinkage.
b. Improved version with a minimum of deformation.
1 = deformation.

Fig. 1.544. Injection moulded gear wheel with metal insert.

Injection moulded wheels should preferably be made with a land flash: this permits a simpler mould, and produces a stronger tooth. However, plastics gear wheels can be injection moulded without a land flash in a good mould.

One or two consequences of using plastics gear wheels should be noted:
(i) Most plastics have a much higher coefficient of expansion than metal. This must be given due consideration for transmissions with metal inserts, exposed to a wide range of temperatures.

(ii) Because the material shrinks after being injection moulded, the mould cavity must be larger than the product. A reasonable approximation of the theoretical tooth shape is obtained on the injection moulded wheel, by giving the mould a modulus

$$m' = \left(1 + \frac{\% \text{ shrinkage}}{100}\right) m.$$

(iii) In the case of plastics sensitive to moisture, it is necessary to allow some deviation from the true dimensions, and the correct theoretical tooth shape.

(iv) Some deformation due to uneven shrinkage is possible (Fig. 1.543).

These conditions can be covered by:

● Employing uniform wall thicknesses wherever possible.
● Keeping the shape symmetrical.
● Using a metal insert (Fig. 1.544).
● Moulding the product symmetrically.

(c) *Gear wheels and pinions combined*

In reduction gears involving several transmissions, gear wheels and pinions are combined (Figs. 1.545 to 1.551). They can be united in a variety of ways. The joint must keep the relative wobble of the two wheels within specified limits (tight limits where the module is small), whilst transmitting the required torque.

Pinion and wheel can be fastened individually on a common shaft. The two gears are usually joined direct by:

● Pressing in, spinning over, riveting, etc.
● Sintering or injection moulding in one piece.

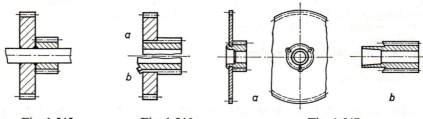

Fig. 1.545   Fig. 1.546   Fig. 1.547

Fig. 1.545. Gear wheel and pinion pressed or glued on a common shaft.
Fig. 1.546. Gear wheel fastened on pinion with one end turned off as far as the
      pitch circle: the tooth roots counteract twisting.
      a. Pressed in.   *b.* Staked.
Fig. 1.547. Gear wheel rotary riveted to hub, with driving torque improved by:
      a. Three indentations.
      b. Points turned on teeth of softer wheels (plastics).

Larger wheels can be fixed by screws (detachable), tubular rivets or solid rivets, provided they are centred accurately in relation to one another. Sintered wheels, impregnated with oil, constitute a self-lubricating bearing on a fixed shaft.

Arrangements of gear wheels in mesh are many and varied. Fig. 1.552 is an example of a transmission used in instruments, particularly with accelerated drives. Here, thin rolled journals are often employed: a steady supply of lubricating oil from chambers in the cover plate ensures long life. Jewel bearings are often used in high-grade mechanisms or movements (see also Volume 3, Chapter 4).

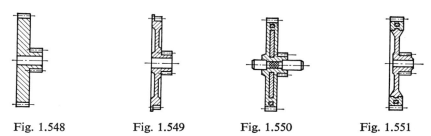

Fig. 1.548   Fig. 1.549   Fig. 1.550   Fig. 1.551

Fig. 1.548. Sintered combination of gear wheel and pinion, also suitable for self-lubrication on stationary shaft (after being impregnated with oil).

Fig. 1.549. Injection moulded gear-and-pinion combination with land flash.

Fig. 1.550. Injection moulded gear-and-pinion combination with metal insert and embedded shaft (screwed in and knurled).

Fig. 1.551. Injection moulded metal pinion as insert in plastic gear ring.

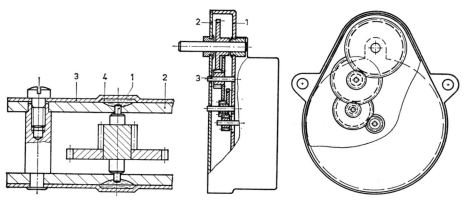

Fig. 1.552. Gear wheel transmission for instruments.
1 = thin rolled journal.
2 = plate.
3 = cover plate.
4 = oil chamber.

Fig. 1.553. Reduction gear suitable for mass production.
1 = housing.
2 = cover plate.
3 = stationary shaft.

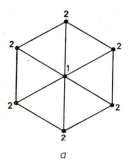

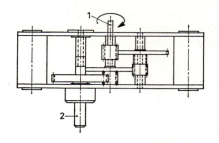

a

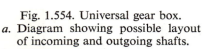

Fig. 1.554. Universal gear box.
a. Diagram showing possible layout
of incoming and outgoing shafts.
b. Actual construction.
     1 = incoming shaft.
     2 = outgoing shaft.

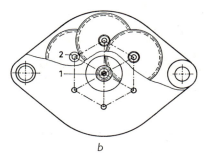

b

Fig. 1.553 shows a version suitable for mass production. Reduction gear boxes of this kind are widely used as programme switch drives in washing machines, etc. Sintered and oil-impregnated gear-and-pinion combinations revolve on hardened, stationary shafts. To avoid an expensive interference fit of shaft in plate, the shaft can be used as a tool (Fig. 1.273*d* and *e*) to punch its own seat in the housing: this ensures a sufficiently tight fit. The shaft may be permitted to fit more loosely in the cover plate. The plate can be rotary riveted, glued or soldered on to the reduction gear box.

With several shafts in, say, steel inserts, cheap plastics bearings moulded-in by one operation can often be used (Volume 11, Chapter 5).

For universal gearboxes, which have to accommodate various transmission ratios, a regular pattern of holes (Fig. 1.554*a*), suitable for different transmissions with identical centre distances, can often be employed with success. For example: transmission ratios of 1/1, 1/2, 1/3, 1/4, 1/5 with numbers of teeth on gear wheel/pinion: 30/30, 20/40, 15/45, 12/48, 10/50.

Given the same module, the centre distance is $\dfrac{60\,m}{2}$.

(d) *Transmissions with little or no backlash*[10, 25, 27]

Transmissions with little backlash and high efficiency can be constructed with precision wheels. Of course, there must also be very little variation in the centre distances and bearings. Such transmissions can be realized in several ways.

In tactile instruments, or probes, as in clocks, backlash is eliminated from a gear train by the pressure of a spring wound up into a flat spiral (Fig. 1.555). Motion is transmitted without backlash in both directions between probe and pointer-spindle.

Backlash can be eliminated from a single transmission by employing a pair of gear wheels instead of a single wheel, one fixed on the shaft, the other free to rotate on it. A tension or compression spring ensures that the detached wheel can also transmit a torque greater than the driving torque (Figs. 1.556a and b). A drawback of the compression spring is that it tends to relax by bending into an arc, thereby causing extra friction when the wheels move in relation to one another. If the angle of rotation of the gear wheels is smaller than 360°, a bent flat spring can be used instead (Fig. 1.556c).

Another possible method is to employ an injection moulded gear wheel of tough, elastic plastic (Figs. 1.557a and b). By allowing slightly less than the theoretical centre distance between pinion and gear, transmission without backlash is obtained through the elastic deformation of the gear ring. To obtain adequate bottom clearance, despite a certain amount of wear, the crest is often dropped about 0·2 m.

The same effect is obtained with non-elastic wheels by means of an extra spring, to press the teeth of the anti-backlash (involute) transmission into mesh, or by using the elasticity of an existing structural member for the same purpose. This can be done by installing one of the shaft bearings in a lever pulled by a helical tension spring (Fig. 1.557c). Provided the modules are small, the shaft oscillation is minimal and the (short) journal bearing, that is, a round hole in a bracket, is just able to absorb it. The lever pivot must be

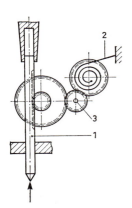

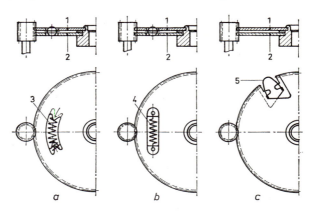

Fig. 1.555. Anti-back-
lash transmission in tac-
tile instrument.
1 = probe.
2 = flat spiral spring.
3 = pointer spindle.

Fig. 1.556. Anti-backlash gear with pair of wheels
(split wheels).
   a. With compression spring.
   b. With tension spring.
   c. With bent flat spring.
   1 = fixed gear wheel.   4 = tension spring.
   2 = detached gear wheel. 5 = bent flat spring.
   3 = compression spring.

positioned as near as possible to the point of application of tooth pressure in both directions of rotation, so that the torque exerted on the lever by tooth pressure does not affect the action of the spring unduly.

The transmission shown later in Fig. 1.569, with two pinions pressed into mesh with two racks, is likewise free from backlash. The two racks are part of a spring slide, and are necessary to guide the slide in a straight line.

An anti-backlash transmission for one direction of rotation appears in Fig. 1.558. It consists of two (involute) gear wheels, one of which has a tooth fewer than the other, connected on the same rolling circle. The wheel governing the transmission is joined to the shaft: the other, with the extra tooth, is pressed against it by a spring. Rotation sets up a frictional torque, which presses the tooth faces of the second wheel against the pinion tooth faces which are not under load.

An anti-backlash gearbox is shown schematically in Fig. 1.559. The complete transmission is duplicated and driven by a single pinion on the ingoing shaft. One of the spur wheels on the outgoing shaft is joined to the

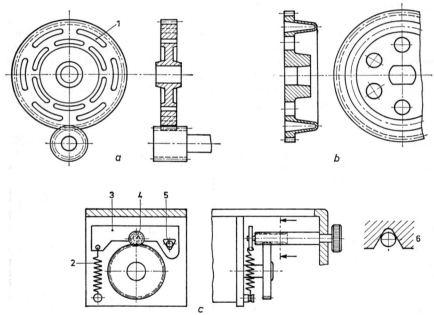

Fig. 1.557. Anti-backlash gears obtained by pressing the teeth of wheel and pinion into mesh.
a. With flexible gear wheel by kidney-shaped slots.
b. With flexible gear wheel by circular gutter.
c. With tension spring and lever structure.
1 = kidney-shaped slots.
2 = tension spring.
3 = lever.
4 = pivot of pinion spindle in lever.
5 = lever pivot.
6 = V-shaped bearings of 4 and 5 to counteract backlash.

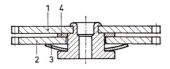

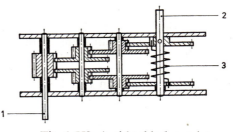

Fig 1.558. Anti-backlash transmission for one direction of rotation.
1 = fixed spur wheel.
2 = free spur wheel with one extra tooth.
3 = coned disc spring.
4 = ring of friction material.

Fig. 1.559. Anti-backlash gearbox.
1 = ingoing shaft.
2 = outgoing shaft.
3 = torsion spring.

shaft: the other is free to rotate on it. The two wheels are linked by a torsion spring, the torque output of which must be greater than the torque required on the outgoing shaft.

1.12.3 Gearing (gear transmissions)[17, 22]

(a) Of small dimensions[18, 20, 26, 32]

The dimensions of gearboxes (e.g. for motor drives) with substantial reductions are much governed by the minimum number of teeth on the pinions, the module, and the reduction per individual transmission. In the case of (involute) spur gears with the theoretical basic rack profile, positively corrected pinions having not less than seven teeth can be milled without undercutting. But, because of tolerances on pinion, mating wheel, bearing and centre distance, it is virtually impossible to combine them into a smooth running transmission.

The minimum number of pinion teeth is smaller for an ordinary (involute) helical gear (see next section). It also depends on the tooth angle. By reducing the addendum, it is even possible to make pinions with one tooth which mesh with wheels of sufficient width to provide a smooth-running transmission, with an efficiency comparable to that of a spur gearbox having the same transmission ratio. The overall size of the gear can be reduced very much more by employing these pinions, but their smaller load capacity is a drawback.

As seen in Sections 1.12.1 (a) and (b), straight cycloidal transmissions, having pinions with six teeth, are used in clock mechanisms.

(b) Parallel shafts and helical teeth

The advantages and disadvantages of helical gearing, compared with spur gearing, are as follows:
- Quieter running, less noise.
- Stronger teeth.
- Minimum number of teeth on pinion is appreciably smaller.
- Axial forces are involved.

● The wheels cannot be used as change wheels.

The axial forces can be eliminated by employing herringbone teeth, by fitting two wheels with identical dimensions but with their teeth angled in opposite directions on the same shaft.

An example of a gear transmission with helical teeth designed for mass production appears in Fig. 1.560. The overall reduction is 1/375, made up of transmission stages 3/45 and 2/50. The three-toothed pinion and the gear wheels are injection moulded in nylon, and the pinion with two teeth is obtained by cold forming or cold working, which involves a shorter run-out than milling and therefore permits a more compact structure. Because of the difference in coefficient of expansion between the steel plates and the wheels, the latter are provided with inserts. This enables the reduction gear to be operated throughout a wide range of temperature. The three plastic bearings are moulded in the plates in one operation. The dimensions are about 30% smaller than those of a spur gearbox with the same transmission ratio. At the same time, the overall efficiency of a spur gear transmission of the same size is slightly higher.

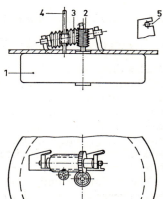

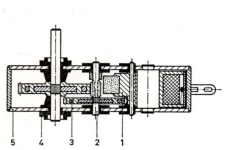

Fig. 1.560. Synchronous motor integrated with reduction gear with helical teeth, suitable for mass production.
1 = rotor with three-toothed pinion.
2 = shaft with pinion with two teeth.
3 = insert in gear wheel.
4 = plastics bearing.
5 = plate.

Fig. 1.561. Worm wheel transmission for clock mechanism.
1 = motor.
2 = motor worm.
3 = second worm.
4 = spindle of seconds hand.
5 = bearing.

(c) *Crossed shafts*

In precision engineering, helical and worm gear transmissions are mainly used for this purpose. Their efficiency is generally somewhat lower than for the gears discussed so far, but they are quieter-running. If the worm wheel is not made too wide, an ordinary spur gear wheel serves the purpose: the worm shaft must then be slanted at an angle equal to the pitch angle. Figure 1.561 shows two worm wheel transmissions with spur gear wheels

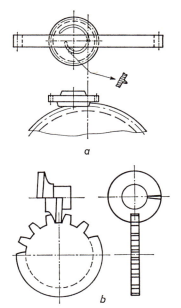

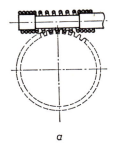

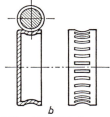

a

b

Fig. 1.562. Simplified worm - and - wormwheel construction.

a. Injection moulded worm with spiral tooth, meshing with injection moulded worm wheel having special teeth.

b. Worm pressed from sheet material meshing with blanked spur gear having conjugate teeth.

Fig. 1.563. Simplified worm - and - wormwheel construction.

a. Helical spring as worm, meshing with worm wheel having conjugate teeth.

b. Worm with normal screw thread meshing with worm wheel, pressed from sheet material, having conjugate teeth.

used as worm wheels, as is often done in clock mechanisms. The motor worm and the spindle of the seconds hand retain the second worm, supported in elongated holes in the plate tabs.

When injection moulded worms are employed, the unscrewing of the worm from the mould necessitates a long moulding cycle. With a seam through the centre-line of the worm, there is risk of flash on the worm tooth.

Figure 1.562*a* shows a gear with a worm that can be injection moulded without difficulty. The smallest radius of curvature of the worm must not be too small, nor the worm wheel too wide, otherwise the worm will jam. The worm wheel must be given either straight or slanting teeth, depending on the position of the worm axis. Another greatly simplified construction is shown in Fig. 1.562*b*, where the worm has only one turn and is punched out of sheet material.

Worms in the form of a helical spring clamped on a shaft are used in the manufacture of toys (Fig. 1.563*a*). The worm wheel is adapted (conjugate) to them. The worm in Fig. 1.563*b* is simply an ordinary screw thread. The wheel is pressed out of sheet material and its teeth made to match the worm as far as possible.

(d) *Intersecting shafts*[29]

Where shafts intersect, bevel gears and crown-wheel-and-pinion trans-missions are mainly employed. A drawback of bevel gears, particularly in mass production, is that when the teeth are straight, the apices of the rolling cones must coincide, if the gear is to run smoothly. The crown-wheel-and-pinion transmission (Fig. 1.564) is a relatively simple construction, but its running characteristics are inferior to those of a bevel gear transmission. Crown-wheels are made by index-milling, -cutting, or blanking and drawing. The last method consists of blanking a flat gear wheel with a tooth shape established by experiment and then raising the teeth by drawing. A reversing mechanism is easily constructed by means of a crown-wheel-and-pinion transmission (Fig. 1.565).

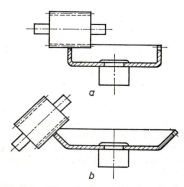

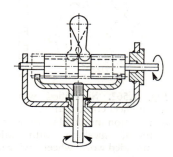

Fig. 1.564. Crown-wheel-and-pinion transmission.
a. Shafts intersect at right-angles.
b. Shafts intersect at 45°.

Fig. 1.565. Reversing mechanism with crown-wheel-and-pinion system.

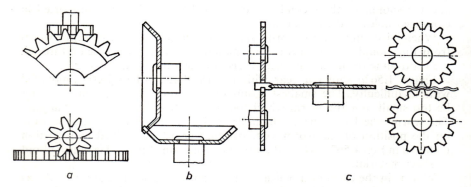

Fig. 1.566. Right-angle transmission with straight-toothed (spur) wheels.
a. Ordinary (cycloidal) spur wheels.   *c*. Gear wheel with corrugated periphery,
b. Two crown-wheels.   driven between two spur wheels of
                                                                      matched tooth shape.

A right-angle transmission suitable for light loads (Fig. 1.566*a*) can also be made with ordinary spur gear wheels. Other simplified transmissions for intersecting shafts are shown in Figs. 1.566*b* and *c*.

1.12.4 Special transmissions

(a) *Racks*

By means of a rack-and-pinion, rotary motion can be converted into translation, or vice versa. For example, Fig. 1.567 illustrates the principle as applied to the adjustment of a stand.

It sometimes happens that the translational member also has to rotate on its longitudinal axis, e.g. in an automatic tuner for television reception (Fig. 1.568). Here, a shaft, with parallel grooves turned in it to form a standard tooth profile, is used as the rack. Where a less accurate transmission will

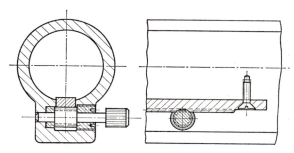

Fig. 1.567. Rack-and-pinion drive used to adjust stand.

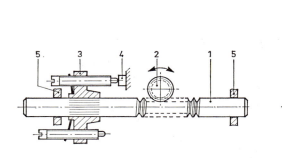

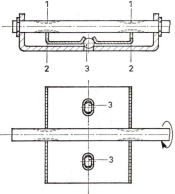

Fig. 1.568. Part of automatic tuner with rack-and-pinion involving rotary rack (or lead screw).
1 = rack. 4 = stop.
2 = pinion. 5 = bearings.
3 = turret.

Fig. 1.569. Double rack-and-pinion system for anti-backlash transmission.
1 = pinion. 3 = guide ball.
2 = rack.

suffice, a combination of a trapezoidal screw and a pinion with helical teeth can be employed.

Fig. 1.569 shows an anti-backlash transmission suitable for cheap mass production. The double rack is blanked from thin sheet material and the two guide balls preloaded on assembly.

(b) *High transmission ratios*[23, 28, 32, 33]

Substantial reductions can be obtained within a small compass by means of special (involute) planetary transmissions. They are generally less efficient than the spur gearboxes previously discussed (see also Volume 1, Section 2.8). Fig. 1.570 is an example of an injection moulded planetary gear with a transmission ratio of 1/10 000. The number of teeth on the different wheels is indicated in the diagram. To avoid bad meshing, a profile with a large pressure-angle is chosen.

The planetary transmission with external teeth (Fig. 1.571) represents another possible version. A special construction contains an harmonic drive, in which the outer teeth are formed in a sleeve capable of elastic deformation (Fig. 1.572). These teeth unwind on the inner teeth of the stationary housing, thereby producing the motion of a planet wheel.

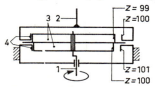

Fig. 1.570. Injection moulded planetary gear with internal teeth and a high transmission ratio.
1 = ingoing shaft.
2 = outgoing shaft.
3 = planet wheels.
4 = internal teeth.

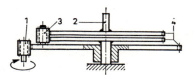

Fig. 1.571. Planetary transmission with external teeth and a high transmission ratio.
1 = ingoing shaft.
2 = outgoing shaft.
3 = planet wheel.
4 = external teeth.

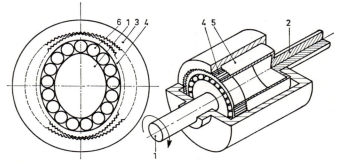

Fig. 1.572. Harmonic drive: simple (plastics) structure for high transmission ratio.
1 = ingoing shaft.
2 = outgoing shaft.
3 = stationary housing.
4 = elastic gear ring.
5 = bush to unite gear-ring with outgoing shaft.
6 = integral ball bearing.

REFERENCES

[1] RICHTER, V. VOSS and KOZER, *Bauelemente der Feinmechanik*, Verlag Technik, Berlin, 1954.
[2] K. HAIN, *Die Feinwerktechnik*, Fachbuchverlag Dr. Pfanneberg, Giessen, 1953.
[3] G. SCHLEE, *Feinmechanische Bauteile*, Verlag Konrad Wittwer, Stuttgart, 1950.
[4] K. RABE, *Grundlagen Feinmechanischer Konstruktionen*, Anton Ziemsen Verlag, Wittenberg/Lutherstadt, 1942.
[5] A. J. DONKERSLOOT, *Tandwielen*, Delftse Uitgeversmaatschappij, Delft, 1963.
[6] F. WOLF, *Die Zahnrädergetriebe in der Feinwerktechnik*, Deutscher Fachzeitschriften- und Fachbuchverlag, Stuttgart, 1953.
[7] R. REUTEBUCH, *Der Uhrmacher*, W. Kempter, Ulm, 1951.
[8] W. TRYLINSKI, *Uhrwerkverzahnungen*, Feingerätetechnik, 1954, 7, p. 319; 1959, 11, p. 526.
[9] H. JENDRITZKI, *Moderne uurwerkreparatie*, Heisterkamp, Amsterdam, 1955.
[10] 18 *Ways to Control Backlash in Gearing*, Product Engineering, 1959, October 26, p. 71.
[11] RICH, 15 *Ways to Fasten Gears to Shafts*, Product Engineering, 1960, May 30, p. 43.
[12] G. NIEMANN and K. ROTH, *Zahnformen und Getriebeeigenschaften bei Verzahnungen der Feinwerktechnik*, Feinwerktechnik, 1964, 9, p. 344; 10, p. 409 and 12, p. 538.
[13] G. HILDEBRAND, *Verzahnungen der Feinwerktechnik*, Feinwerktechnik, 1965, 6, p. 278.
[14] *Mechanical Drives Reference Issue*, Machine Design, 1967, September 21.
[15] *Cilindrische tandwielen*, publication VMO9 of F.M.E., Stam, Culemborg.
[16] THOEN, *Enlarged tooth pinions*, Product Engineering, 1961, November 13, p. 113.
[17] *Seven Rules Simplify Instrument Gear Specification*, Product Engineering, 1957, November 25, p. 80.
[18] R. L. THOEN, *Modified Pinions, Fine Pitch Spur-Gears*. A presentation at AGMA's 1961 semi-annual meeting. AGMA 379–02, 1961, October.
[19] *Cilindrische tandwielen met evolvente tanden*, NEN 1628 ($m < 1$ mm), NEN 1629 ($m = 1$ to 20 mm), NEN 1630 (modulussen).
[20] K. ROTH, *Evolventenverzahnungen für parallele Achsen mit Ritselzähnezahlen von 1 bis 7*, VDI-Z, 1965, 6, p. 275.
[21] H. SACHSE, *Der Weg zu einem Einheitsbezugsprofil in der Feingerätetechnik*, Feingerätetechnik, 1961, p. 177.
[22] K. ROTH, *Kennzeichnende Merkmale feinwerktechnischer Konstruktionen*, VDI-Z, 1963, 22, p. 1017; 23, p. 1125.
[23] L. A. GRAHAM, *Planetary Transmissions*, Machine Design, 1946, November, p. 115.
[24] H. KÖHLER, *Übersetzungsverhältnis, Überdeckungsgrad und Verteilung des Eingriffs bei den Verzahnungen der Feinmechanik*, Feingerätetechnik, 1956, 4, p. 156.
[25] V. S. STARODUBOV, *Reduction Gears without Backlash for Numerically Controlled Machine Tools*, Russian Engineering Journal, **46**, 3, p. 16.
[26] THEON, *High Grade Fine-Pitch Gearing*, Machine Design, 1961, January 19, p. 154.
[27] M. S. BRUNO, *Checking Backlash in Gears*, American Machinist/Metalworking Manufacturing, 1963, May 27, p. 77.
[28] MICHALEC, *Precision Gear Trains*, Machine Design, 1966, February 3, p. 126.
[29] R. NAVILLE, *Kegelräder für die Feinwerktechnik*, Feinwerktechnik, 1967, 8, p. 358.

[30] G. NIEMANN and M. UNTERBERGER, *Geräuschminderung bei Zahnrädern*, VDI-Z, 1959, p. 213.

[31] W. TRYLINSKI, *Kraftmomentschwankungen der evolventen Verzahnungen*, Feingerätetechnik, 1962, 4, p. 146.

[32] W. O. DAVIS, *Gears for Small Mechanisms*, Waterlow and Sons, London, 1953.

[33] N. CHIRONIS, *The Harmonic Drive*, Product Engineering, 1960, February 8, p. 47.

[34] S. HILDEBRAND, *Feinmechanische Bauelemente*, 1968, Carl Hanser Verlag, Munich.

1.13 Brakes and dampers (or shock absorbers)[1 – 3, 5]

To slow down or halt the motion of a member, the kinetic energy can be (mainly) converted into heat in the form of frictional losses or eddy-current losses. Continuous forward motion is usually retarded by friction between mechanical members. On the other hand, hydraulic friction, pneumatic friction or eddy-current losses are only employed to suppress oscillation or vibration of a structural member about a position of equilibrium. This is called damping and the constructional elements concerned are called dampers.

1.13.1 Brakes

They are of little importance for rectilinear motion in precision engineering but axially or radially engaging block or band brakes are used to halt rotary motion. In the block brake, several blocks or shoes can be arranged on the periphery of the brake pulley (drum or disc) to relieve the load on the shaft bearing. In precision engineering, the braking forces are usually so small that the bearings can cushion them easily, so that braking on one side only is sufficient. Band brakes are hardly ever used here, but, for particulars, see conventional mechanical engineering literature. Nor is there much to be said about the block brake systems employed.

However, electromagnetically-operated brakes are worth mentioning. They usually work axially and are constructed in the same way as the electromagnetic friction clutches discussed in Section 1.9.10 (c).

The choice of friction material depends above all on the coefficient of friction, the behaviour under varying operating conditions and the amount of wear. Wood, leather, fibre or synthetic resin-bonded paper are used on metal, where a great deal of friction is required. Limited friction and a uniform braking action are best obtained through metal-on-metal, but the system should be so constructed that the metals chosen do not weld (seize) together as a result of the braking forces. For further particulars of friction materials, see Section 1.9.8.

1.13.2 Dampers[4]

The dampers discussed here are used mainly in weighing machines, measuring or test apparatus and so on. The damper supplies a force, opposed to the motion, which increases with the speed of the motion. The energy of the reciprocal motion is exhausted by the damper, which thus restores the state of equilibrium. This is possible only if the energy transmitted by the damper to the moving system in the stationary state is zero, as it is with frictional resistance and eddy-currents.

The degree of damping in the system should be suited to the particular purpose. Where a pointer setting has to be read direct, creeping (aperiodic) travel to a new setting is deprecated, for it is likely to cause misreading. It is better to have the pointer overshoot its mark two or three times before coming to rest there. On the other hand, aperiodic damping is recommended for graphic (or recording) instruments. See also Volume 1, Section 2.4.2.

Hydraulic dampers, or dash-pots, are used where near-aperiodic damping is required of large masses or fast-moving instrument parts. The damping medium is usually a viscous liquid, such as oil or glycerine. The oil must not become gummy or resinous when exposed to air, and must not attack neighbouring materials: also, its viscosity should not vary too much with differences in temperature. Towards lower temperatures, damping increases and the reading may become aperiodic and uncertain. Excessive damping can be countered by (variable) electric heating or by (temporarily) increasing the flow of liquid, sometimes in response to control by a bimetallic strip in the structure.

Fig. 1.573 shows a practical damper for rectilinear or near-rectilinear motion in a weighing machine. The piston (1), movable on the piston rod, contains a number of ports (2). The passage of damping liquid, and therefore the damping effect, is controlled by moving piston (1) in relation to the fixed piston (3). This can be done by turning the knurled nut (4) and thus moving piston (1) via tube (5) against the pressure of spring (6). The liquid (7) does not fill the cylinder completely. For safe transport, the cylinder can be sealed off by screwing the knurled nut into the cylinder cover, which prevents the oil escaping.

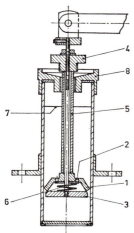

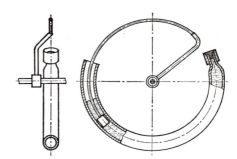

Fig. 1.573. Variable dash-pot with piston for rectilinear motion.
1 = movable piston.
2 = ports.
3 = fixed piston.
4 = knurled nut.
5 = tube.
6 = spring.
7 = liquid.
8 = cylinder cover.

Fig. 1.574. Dash-pot with piston for rotary motion.

Measuring instruments rotating in bearings with a wide angle of deflection can be damped by annular cylinders filled with liquid (Fig. 1.574). Practical dampers of this kind have cylinders with an inside diameter of 4 mm and pistons of 3·2 mm diameter. When the cylinder is made of glass, it can be blown to different inside diameters at different points, so as to vary the damping effect locally. Thus, the damping can be reduced by an expansion coinciding with a stable position of the piston, giving added assurance that it will register correctly.

Despite the effectiveness of hydraulic dampers, air dampers are often preferred, when the presence of liquid in an instrument cannot be tolerated. Air dampers, with a swivelling vane or diaphragm, are usually employed for accurate measuring instruments (Figs. 1.575 and 1.576). The damping effect of the system in Fig. 1.575 is influenced adversely by the gap through which the swivel arm of the vane must pass. In the other system, this is avoided because there is only a small gap to admit the shaft. Fig. 1.577 shows another practical version. The hairpin driving rod may be too weak. The mass moment of inertia of the pivoting vanes is limited by making them out of thin sheet aluminium, strengthened by flanging the edges. On no account, of course, must a vane come in contact with the wall.

Eddy-current dampers operate on the principle that current is induced in a metal plate when it is moved between the poles of a magnet, at right-angles to the field. In doing so, the magnetic field exerts on the plate a force that is

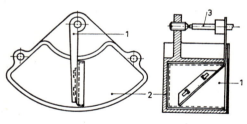

Fig. 1.575. Air damper with vane pivoting on a spindle outside the damping chamber.
    1 = moving vane.
    2 = damping chamber.
    3 = spindle.

Fig. 1.576. Air damper with vane pivoting on a spindle inside the damping chamber.
    1 = moving vane.
    2 = damping chamber.
    3 = spindle.

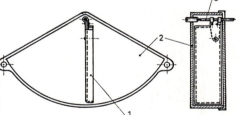

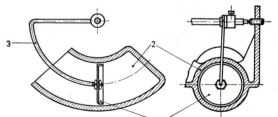

Fig. 1.577. Air damper with round piston and curved damping chamber.
    1 = piston.
    2 = damping chamber.
    3 = driving rod.

proportional to the speed; is inversely proportional to the electrical resistance of the plate material; and depends on the square of the magnetic flux.

Because the effect at rest is nil, the system can function as a damper. The types shown in Fig. 1.578 have aluminium plates, a material that is sufficiently conductive and, at the same time, light enough to have a low mass moment of inertia. A variable magnetic shunt to control the damping effect is possible, but is not shown in the diagram. Eddy-current dampers are used for small precision measuring instruments in which the mass forces to be damped are very small. The featherweight version of Fig. 1.578b is part of a hot-wire instrument and brings about virtually aperiodic damping. Similar dampers are found in torsion spring balances.

The coil former of a moving-coil can be exploited as a damper by making it of aluminium (Fig. 1.579).

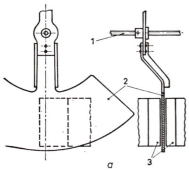
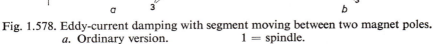

Fig. 1.578. Eddy-current damping with segment moving between two magnet poles.
    a. Ordinary version.   1 = spindle.
    b. Featherweight version.   2 = aluminium segment.
                                3 = permanent magnet.

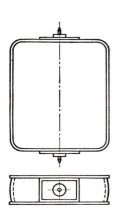

Fig. 1.579. Moving-coil with aluminium frame as coil former and two glued pivots.

REFERENCES

[1] *Mechanical Drives Reference Issue*, Machine Design, 1967, September 21.
[2] RICHTER, V. VOSS and KOZER, *Bauelemente der Feinmechanik*, Verlag Technik, Berlin, 1954.
[3] G. SCHLEE, *Feinmechanische Bauteile*, Verlag Konrad Wittwer, Stuttgart, 1950.
[4] G. MUTZ, *Dämpfung von Prellungen durch einseitige Impulstranslation*, Feinwerktechnik, 1966, 12, p. 581.
[5] S. HILDEBRAND, *Feinmechanische Bauelemente*, 1968, Carl Hanser Verlag, Munich.

1.14 Governors[1–4, 16]

1.14.1 Introduction

The speed of a mechanism in motion can be kept constant by continuous adjustment of the energy supplied to the energy required, or through continuous (or step-by-step) dissipation of the surplus energy by braking. The speed of fairly large machines and apparatus is governed by changing the energy supply, which ensures a favourable level of efficiency. The energy required in precision engineering is so low, however, that it need not be used as sparingly (except in the case of clocks, etc.), so quite simple governors can be employed.

To keep the speed of a movement (usually rotary, like a drive) within given limits, the initial driving torque is set higher than the actual maximum required. How much extra torque is adopted depends on the estimated fall-off in efficiency due to wear, dirt and changes caused by temperature, environment and so on. The surplus torque can be eliminated continuously by a frictional moment. Friction can be generated between solid bodies or through fluid friction or air friction, or it can be superseded by eddy-currents, and afterwards exhausted in electrical conductors.

Intermittent governors block the travel of a moving mechanism by alternately detaining and releasing a driven structural member. In governors of the escapement type, the process has a specific rhythm, in which the driving torque can only move the mechanism step-by-step.

1.14.2 Braking governors[5–8]

These continuously-operating devices eliminate the difference between the energy supplied to a moving mechanism and the energy taken from it. Formulated:

$$E_t = E_a + E_r$$

where E_t is the input energy.
E_a is the output energy.
E_r is the braking energy.

Referred to an equivalent revolving shaft, this formula becomes

$$M_t = M_a + M_r$$

where M_t is the input torque
$\qquad M_a$ is the output torque
$\qquad M_r$ is the braking torque.

The input energy is often obtained from electric motors, springs or weights, and this determines the behaviour of the driving torque as a function of the angular velocity (ω) of the shaft. The relationship between the torque output and this angular velocity may vary, and depends on the nature and purpose of the mechanism. The governor must absorb the torque difference $M_t - M_a$ so as to keep the associated angular velocity within acceptable limits.

This is illustrated by Fig. 1.580, in which the torques M_t, M_a, M_r and $M_a + M_r$ are plotted against ω (solid lines). The magnitude of the (motor) torque M_t is influenced in this case by the battery voltage, and is limited by the two dotted curves on either side of the M_t curve. Other motors have other curves. The torque output M_a is affected by the temperature of the lubricating oil, and may therefore vary between the two straight dotted lines on either side of the straight M_a-line. The sum torque $M_a + M_r$ has as its limits the dotted curves on either side of the $(M_a + M_r)$ curve. To simplify matters it is assumed that the curve M_r, that is, the braking characteristic, does not change.

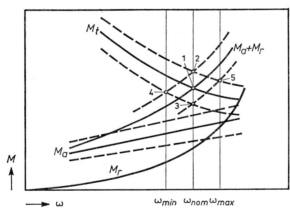

Fig. 1.580. Torque in a braking governor plotted against angular velocity.
M_t = applied torque. M_a = output torque. M_r = braking torque.

With nominal torque values the curves M_t and $M_a + M_r$ intersect at the working point (1): in other words the mechanism stabilizes at the speed ω_{nom}. With maximum and minimum torque, these working points lie at (2) and (3) respectively. In a cold environment and in operation with under voltage the working point is at (4) with a speed ω_{min}: in a hot environment, likewise with excess voltage, the working point is at (5) with a speed ω_{max}. The braking governor keeps the speed variation between the limits ω_{min} and ω_{max}. In given circumstances, this variation decreases as the slope of the braking characteristic becomes steeper, in the region of the working point. For this reason, braking governors can usually be adjusted to optimize the shape and/or position of the characteristic.

(a) *Governors operating with friction between solid bodies*

It follows that governors of this type have to generate a braking force that increases with angular velocity. The obvious course is to utilize the centripetal force of one or more masses rotating with a shaft. This is called a centrifugal governor. As the shaft revolves, a mass (Fig. 1.581) swings out against the force of a spring until, at a given speed (ω_0), the brake shoe approaches the friction track.

With a further increase in speed, the friction track has to hold the mass in its path, whereupon a normal force is set up to generate the braking friction (Fig. 1.582). Fig. 1.583 shows the braking characteristic without the spring. Here, the braking torque is $K_1 . m\omega^2 r . \mu . R$, whereas in the situation of Figs. 1.581 and 1.582 this torque is $(K_1 . m\omega^2 r - K_2 . F) . \mu . R$ where

m is the rotating mass

r is the distance from the centre of gravity of the mass to the axis of rotation.

μ is the coefficient of friction of the brake shoe on the friction track.

R is the distance from friction track to axis of rotation.

K_1 is a factor depending on the suspension of the mass.

K_2 is a factor depending on the arrangement of the spring.

F is the force of the spring at $\omega \geqslant \omega_0$.

$\omega_0 = \sqrt{\dfrac{K_2 . F}{K_1 . m . r}}$ is the velocity at which the mass touches the friction track.

Fig. 1.581. Principle of the centrifugal governor.

        1 = mass.
        2 = friction track.
        3 = spring.

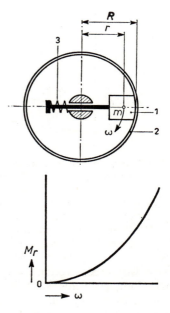

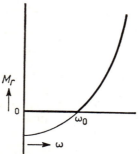

Fig. 1.582. Braking characteristic of a spring-loaded centrifugal governor.

Fig. 1.583. Braking characteristic of a centrifugal governor without spring.

There are two distinct types, one of which acts radially, the other axially.

Fig. 1.584 shows a radial-action centrifugal governor with a fixed friction track and with the masses on individual springs. It can be adjusted to a given speed by varying the length of the springs with the aid of the movable disc (5). This changes the position of the braking characteristic and hence ω_0. The adjustment is carried out with the governor at rest.

Fig. 1.585 depicts a very much more robust form of centrifugal radial-action governor. A system of two masses, pivoting at the ends (3) of the driving member, rotates inside the fixed fraction track in the direction indicated by the arrow. The brake-shoes (5) are thrust outwards on to the friction

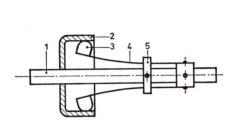

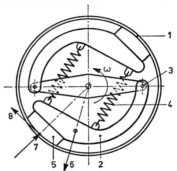

Fig. 1.584. Adjustable centrifugal governor with radial action.

1 = shaft.	4 = spring.
2 = friction track.	5 = movable disc for adjustment.
3 = mass.	

Fig. 1.585. Robust centrifugal governor with radial action.

1 = friction track.	6 = centrifugal force.
2 = mass.	
3 = pivot.	7 = centripetal force.
4 = spring.	
5 = brake-shoe.	8 = friction force.

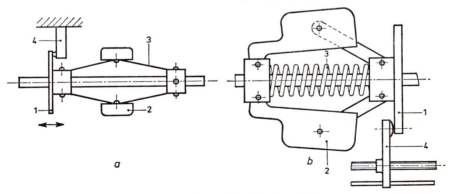

Fig. 1.586. Centrifugal governor with axial action involving movement of friction track.

a. Simple version.
b. Robust version.

1 = friction track.	3 = spring.
2 = mass.	4 = brake-shoe.

track against the force of the springs. The moment of the frictional force (8) opposes the outward swing of the system. In the opposite direction of rotation, the frictional force enhances the moment of the mass at point (3), making the governor stiffer or, in other words, giving it a steeper braking characteristic.

Fig. 1.586 shows two axial-acting governors, a simple version and a more robust construction. In both, the spring force resists the centrifugal force until the friction track is moved far enough to bring it into contact with the brake-shoe. A given speed can be selected by changing the position of the brake-shoe, which can be done with the governor rotating so that the speed can be measured. The friction torque produced in centrifugal governors changes with the coefficient of friction (μ) between the friction surfaces.

Braking characteristics for different values of μ coincide at the speed at which the brake-shoe touches the friction track (angular velocity ω_0 in Fig. 1.582), but diverge when the brake has to absorb more friction velocity.

To limit the effect of changes of friction, the governor should not have to absorb too large a share of the energy supplied.

Fig. 1.587. Centrifugal governor with axial action, incorporating a conical friction
clutch.
    1 = ingoing shaft.
    2 = outgoing shaft.
    3 = mass.
    4 = helical spring.
    5 = friction clutch.

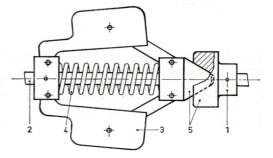

A third type of centrifugal governor, operating on a different principle, is shown in Fig. 1.587. Here, the ingoing shaft (1) is coupled to the remainder of the mechanism by a conical friction clutch (5), and the normal force is supplied by a helical spring (4). When the torque taken from the shaft 2 decreases, the mechanism accelerates and the centrifugal forces grow. This partially compensates the normal force supplied by the spring, so that the friction force generated diminishes and the driving torque eases off, thus slowing the mechanism. The speed stabilizes in such a way that the torque transmitted by the clutch equals the output torque. This friction-clutch governor must not be used where the speed of the shaft responds too abruptly to load changes because this drive is in touch with the changes. However, this objection is overcome when the speed of the driving part is limited separately. The working point can be altered by varying the initial stress of the spring.

Later, in Section 2.3, governors for cine-devices and dialling apparatus are described.

(b) *Fluid-friction braking governors*

Governors employing fluid friction are invariably of the piston type and therefore used mainly to control rectilinear (or almost rectilinear) motion. The braking action is obtained from friction between the individual particles of the fluid, and between these particles and the piston and cylinder walls. The resistance is also enhanced by vortices set up as the fluid passes through the narrow gap between cylinder wall and piston. The magnitude of the braking effect depends on having a suitable shape for this gap (Fig. 1.588). A piston of rectangular cross-section usually offers more resistance than one with rounded edges, and less resistance than a piston with sharp-edged grooves around the periphery.

In most cases, a viscous oil or glycerine is used (see also Section 1.13.2 on hydraulic dampers). Any variation in viscosity can be compensated by varying the bore of a by-pass duct provided for the purpose.

As a rule, the braking effect is the same in both directions. To make it weaker in one direction than the other, one or more (ball) valves can be incorporated in the piston. Governors of this type are used in gas geysers.

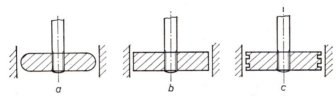

Fig. 1.588. Piston of braking governor operating with fluid friction.

a. With round edges.   *c.* Sharp-edged grooves around periphery.
b. Rectangular section.

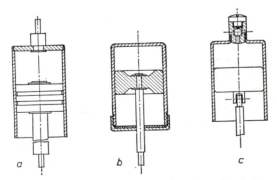

Fig. 1.589. Piston governor operating with air friction.

a. With open cylinder and piston with sharp-edged grooves.
b. With closed cylinder, for dusty environment.
c. With valve for reduced braking action in one direction.

(c) *Air-friction braking governors*

These governors can be divided into two groups, according to the kind of motion to be controlled, namely:

● Piston governors for (approximately) rectilinear motion.
● Wind-vane governors for rotary motion.

The medium in both is invariably air.

Because the viscosity of air is low, very narrow passages are necessary to obtain a good braking action from a piston governor. If the air has to pass from one end of the cylinder along the piston to the other end, a clearance not greater than some hundredths of a millimetre must be provided between piston and cylinder wall. Since, ideally, there should be no extra resistance, and the piston and cylinder walls cannot be lubricated, the coefficient of friction between them has to be very low indeed.

For this reason, the piston is sometimes made of compressed graphite, with sharp-edged grooves around the periphery to add to the gap resistance (Fig. 1.589a). In dusty environments the cylinder should be closed. (Fig. 1.589b). To weaken the braking action in one direction, a ball valve can be included (Fig. 1.589c). Where the stroke is relatively short, the piston can be replaced by a diaphragm: this eliminates piston friction, but the stiffness of the diaphragm is often troublesome.

Air-friction piston governors are rarely used in precision engineering: very much more is made of wind-vane governors. They differ from the pneumatic dampers (Section 1.13.2) for, instead of a single airfoil acting in a sealed chamber, there is a vane rotating freely and at high speed in air, usually on the final shaft of a gear train with a high (accelerated) transmission ratio. The vane is usually made of flat sheet material and the force acting on it is proportional to the area of the vane and the square of the velocity. From this it follows that, where the vane area is kept small, in order to save space, it operates most efficiently at high speed. Speeds up to 3000 r.p.m. are employed. Simple wind-vane governors are rectangular (Fig. 1.590). At relatively high speeds the braking characteristic is rather steep.

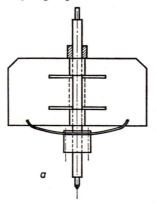
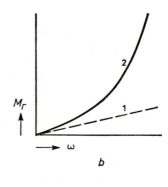

Fig. 1.590. Simple wind-vane governor.

a. Rectangular type.	1 = friction torque of bearing.
b. Braking characteristic.	2 = total torque.

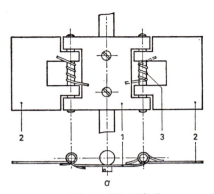
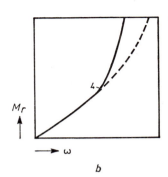

Fig. 1.591. Wind-vane governor with hinged vanes.

a. Opened outwards.   1 = main vane.   3 = spring.
b. Braking characteristic.   2 = vanes opened outwards.   4 = start of outswing.

An appreciably steeper characteristic can be obtained by employing hanged vanes that open outwards (Fig. 1.591). Centrifugal force causes the two vanes to open against the force of torsion springs. It will be seen from Fig. 1.591b that the braking characteristic steepens appreciably after the outward swing of the vanes. It is here that the controlling action is most effective therefore. A similar effect is produced by the vane shown in Fig. 1.592. Towards higher speeds this vane assumes a more horizontal position against the pull of the helical spring, thus greatly extending the

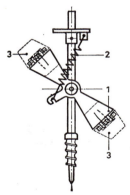
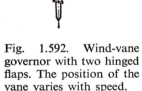
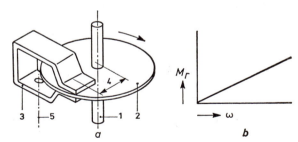

Fig. 1.592. Wind-vane governor with two hinged flaps. The position of the vane varies with speed.
    1 = main vane.
    2 = helical spring.
    3 = hinged flaps.

Fig. 1.593. Braking governor operated by eddy-currents.
    a. View of governor.
    b. Braking characteristic.
    1 = shaft.
    2 = brake disc.
    3 = permanent magnet.
    4 = brake arm.
    5 = adjustable spindle of magnet.

arm on which the resultant air resistance acts. Two hinged flaps (indicated by dotted lines in the diagram) make the control more effective. Used in clock mechanisms and musical boxes.

(d) *Eddy-current braking governors*

Particulars of the braking effect encountered by a highly conductive disc rotating between the poles of a magnet, are given in Section 1.13.2 on eddy-current dampers. The eddy-current governor is a suitable choice for low angular velocities and small torques. A round brake disc (Fig. 1.593a) is fitted on the shaft whose speed is to be regulated, so that it rotates between the poles of a permanent magnet without touching them. The disc is made of a highly conductive material, usually copper or aluminium. Since this disc, in motion across the magnetic flux, is subjected to a braking moment that increases linearly with the angular velocity, governors of this type are employed when the braking moment must be proportional to the speed of rotation (Fig. 1.593b) as, for example, in kilowatt-hour meters.

Because of the braking characteristic, eddy-current governors are less effective than the other types. Since the performance depends partly on the size of the magnetic flux through the disc, the slope of the characteristic can be reduced by a magnetic shunt. The braking torque can be raised by lengthening the brake arm (4): to do so, the magnet is twisted slightly on the shaft (5), by employing a suitable screw fastening.

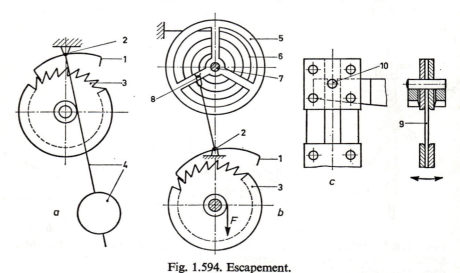

Fig. 1.594. Escapement.

a. With pendulum.	1 = anchor (with pallets).	6 = hairspring.
b. With balance.	2 = anchor staff.	7 = balance staff.
c. Pendulum suspension.	3 = escapement wheel.	8 = pin and fork.
	4 = pendulum.	9 = flat spring
	5 = balance.	bearing.
		10 = pivot.

1.14.3 Escapement devices[9 – 12, 14, 15]

(a) Introduction

Unlike braking governors, escapements do not operate continuously, but engage the moving mechanism periodically, to produce intermittent motion. Since we shall be concerned exclusively with rotary motion, only intermittent motion mechanisms (or escapements) that control rotating parts will be discussed. Such a governor (Fig. 1.594) consists of a reciprocating member (*anchor*), often influenced by a pendulum or balance, and a toothed wheel (the *escapement wheel* or '*scape wheel*) with which the pallets engage. There are many different kinds, but only one example is given.

A pendulum or balance renders the mechanism self-oscillating, which is necessary to keep the motion of the anchor uniform when the mechanism is used to measure time. Such a mechanism is an escapement with its own natural period of oscillation. Where the accuracy of the control has to meet less stringent requirements, the oscillatory effect can be omitted. The speed of the anchor movement of such an escapement device without natural oscillation then depends mainly on the mass moment of inertia of the anchor and the size of the driving force.

With a *pendulum* the clock mechanism is cumbersome to move and the pendulum has to hang vertically. When the deviation from the central position is small, the period of oscillation for a full swing to-and-fro follows from the equation:

$$T = 2\pi \sqrt{\left(\frac{I}{mlg}\right)}$$

where I is the mass moment of inertia,
  m is the mass of the pendulum,
  l is the distance from the centre of gravity of the pendulum to the centre of rotation and
  g is the acceleration due to gravity.

Introducing the reduced pendulum length l_r for I/ml, which defines the period of oscillation, gives:

$$T = 2\pi \sqrt{\frac{l_r}{g}}$$

The only factor, other than the acceleration due to gravity, governing the period of oscillation is the reduced pendulum length, so the period of oscillation can be varied quite simply by adjusting the length. Since the length of the pendulum also depends on temperature, accurate clock movements have to be given a pendulum made of Invar, or a compensation pendulum so constructed that the reduced pendulum length (distance from centre of oscillation to point of suspension) remains constant despite temperature changes. The pendulum is suspended from a spring bearing (Fig. 1.594c). A pivot enables the pendulum to oscillate in a vertical plane.

Mechanisms assuming a variable position and/or inclination are governed by a *balance* or *balance wheel* (Figs. 1.594b and 1.595). This is a kind of flywheel, supported by pivots on its axis of gravity, that oscillates together with a spiral flexural spring (balance spring or hairspring). The period of oscillation is defined by:

$$T = 2\pi \sqrt{\left(\frac{I}{M}\right)}$$

where I is the mass moment of inertia of the flywheel and M is the restoring torque of the hairspring per radian.

The restoring torque is determined by:

$$M = \frac{Ebh^3}{12l}$$

where E is the modulus of elasticity, b is the width, h the thickness and l the length of the rectangular spring material.

The period of oscillation is increased or reduced by adjusting the effective length of the spring. This is done by turning the hairspring key (compass), causing two pins to encompass the spring in a different place (Fig. 1.596). Although there should be some play between hairspring and pins, it is kept to a minimum.

Differences in temperature affect the modulus of elasticity and length of the hairspring and therefore also the period of oscillation. This can be avoided to a large extent by choosing Invar or Elinvar.

More stringent requirements can be imposed on compensation balances, in which the two-part rim of the balance wheel is bimetallic, to compensate the variations of the spiral as much as possible (Fig. 1.597). This system is used in chronometers. Adjusting (quarter) screws enable the axis of gravity to be brought into exact coincidence with the bearing axis.

As mentioned earlier, the object of the escapement is to convert the oscillatory motion of the anchor into intermittent progressive rotary motion. To compensate for the inevitable energy loss involved, energy must be delivered to the anchor to keep it moving. The detention and subsequent release of a tooth on the 'scape wheel must therefore be accomplished in such a way that the wheel also keeps the oscillating members in motion.

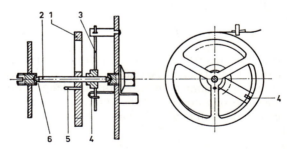

Fig. 1.595. Balance mechanism.

1 = balance.
2 = balance staff.
3 = hairspring.
4 = hairspring key or spanner.

5 = pin to transmit motion between balance mechanism and anchor.
6 = pivot.

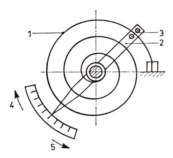

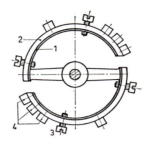

Fig. 1.596. Hairspring key (compass).
1 = hairspring.   4 = slow.
2 = compass needle. 5 = fast.
3 = two pins.

Fig. 1.597. Compensation balance.
1 = steel.   3 = adjusting screws.
2 = brass.   4 = masses.

All the escapements discussed in the following paragraphs are used in self-oscillating mechanisms (clock movements). The last subsection (f) deals with mechanisms that are not self-oscillating.

(b) *Recoil escapements*

Fig. 1.598 illustrates the action of an escapement. The anchor-bills engage the teeth of the 'scape wheel alternately. These teeth are so profiled that the lifting (or impulse) face initiates the movement of the anchor, whilst this in turn stops the motion of the 'scape wheel a moment later by colliding with the braking (or locking) face of a tooth. The tooth profile depends on the shape of the two anchor-bills, which can be designed to make this profile symmetrical. The same (right-hand) tooth face then acts as both impulse face and locking face.

In Fig. 1.598, the left-hand anchor-bill locks the 'scape wheel until the anchor has pivoted far enough in a clockwise direction to allow the locked tooth to pass (Fig. 1.598b). In the meantime, the right-hand anchor-bill has reached a point between two teeth of the wheel and intercepts this again

Fig. 1.598. Action of a recoil escapement.
a. Left-hand pallet locks 'scape wheel.   1 = anchor.
b. Left-hand pallet releases 'scape wheel.   2 = 'scape wheel.
c. Right-hand pallet locks 'scape wheel.   3 = anchor arbor.
d. Right-hand pallet releases 'scape wheel.   4 = impulse face.
e. Left-hand pallet locks 'scape wheel.   5 = locking face.
                                                                  6 = left-hand anchor-bill (entrance).
                                                                  7 = right-hand anchor-bill (exit).

(Fig. 1.598c). The impact of the wheel tooth against the anchor-bill produces an audible tick. The anchor then pivots slightly further to the right, so that the right-hand anchor-bill forces the 'scape wheel back a little, whereupon the movement reverses and the 'scape wheel propels the anchor to the position indicated in Fig. 1.598d, at which the locked tooth can pass to the situation of Fig. 1.598e, and the clock ticks again.

The left-hand side of the anchor is called the entrance and the right-hand side the exit of the anchor. Between the positions illustrated in Fig. 1.598b and c, and between those shown in Fig. 1.598d and e, the wheel makes a free stroke. This is necessary to take up the spread in the size of the parts.

The escapement described here is of the recoil type: the 'scape wheel (and therefore also the seconds hand, if any) recoils slightly after each tick of the clock. The recoil involves friction losses and wear, whilst changes in the motive force produce little change in the angle of oscillation of the pendulum. Recoil escapements were widely used in the past, but nowadays are found only in spring-driven clocks and in cheaper clocks with other sources of power.

The recoil escapement occurs in various forms. Fig. 1.599 shows two practical versions with a hardened steel nib anchor for a short pendulum, as used in a pendulum clock, and a nib anchor of bent and hardened strip for cheap cuckoo clocks. The 'scape wheel is usually of hard brass.

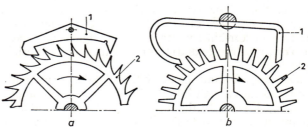

Fig. 1.599. Recoil escapement with nib anchor.

a. Massive nib anchor.   1 = nib anchor.
b. Bent nib anchor.   2 = 'scape wheel.

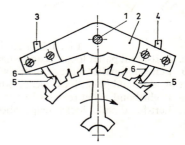

Fig. 1.600. Graham-escapement.

1 = anchor arbor.   4 = exit pallet.
2 = anchor.   5 = impulse face.
3 = entrance pallet.   6 = locking face.

(c) Dead-beat and frictional-rest escapements

The *Graham-escapement* (Fig. 1.600) is constructed to minimize the impulse arc, or that portion of the anchor stroke during which it is impelled by the 'scape wheel. This impulse is delivered near the central position of the anchor: in the other positions the anchor-bills move in and out of the 'scape wheel as far as possible unhindered. The anchor-bills, which are separate parts here, are called pallets. The motive force supplied by the 'scape wheel is transmitted to the anchor intermittently via the impulse faces on the entrance and exit pallets: the pulses occur as the anchor is passing its central position. This escapement is known as "dead-beat". Owing to the shape of the rest or locking faces, which are on cylindrical surfaces about the anchor axis, the 'scape wheel remains at rest, instead of recoiling, during the forward motion of the anchor after the clock ticks. This is, in fact, its principal advantage compared with recoil escapements. The anchor is made of solid steel or brass, with hardened, adjustable steel pallets engaging with a hard brass 'scape wheel. The escapement is employed in large pendulum clocks.

Of a similar type, called frictional-rest, is the *cylinder escapement* shown in Fig. 1.601*a*. This may be considered as derived from the Graham-escapement by bringing the teeth together until they span only one tooth. At the cylinder, the 'scape wheel teeth transmit an angular momentum to the reciprocating anchor via the slanting impulse face on the top of the tooth. It will be seen from the diagram that the tooth is detained in the same way as in the Graham-escapement. The inner and outer sides of the cylinder are rest faces. The friction between cylinder and tooth during the reciprocating motion sets up frictional resistance which is difficult to control. Fig. 1.601*b* is the longitudinal section of a cylinder as used in a pocket watch. This design is no longer used very much because it is difficult to regulate.

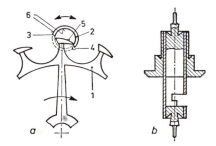

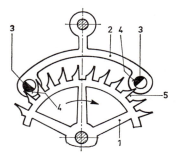

Fig. 1.601. Cylinder escapement.
a. Escapement.
b. Longitudinal section of cylinder of
   a pocket watch.
1 = 'scape wheel.　4 = exit lip.
2 = cylinder.　　　5 = impulse face.
3 = entrance lip.　6 = rest, or
                      locking face.

Fig. 1.602. Escapement with Brocot
           pin-pallet.
1 = 'scape wheel.　4 = impulse face.
2 = anchor.　　　　5 = rest face.
3 = pallet pads.

In high-grade movements for clocks, an escapement with a *Brocot* pin-pallet is employed (Fig. 1.602). It functions in very much the same way as the Graham-escapement. Here, the flank of the 'scape wheel tooth forms the locking, or rest face. This is really only approximately correct, however, since, for true dead-beat, the tooth flank must be concave at the entrance and convex at the exit, which cannot be done. The impulse is transmitted by the tip of the tooth resting on the semi-circular side of the D-shaped pallet. These pallets are generally made of ruby or sapphire. This is the cheapest form of escapement with jewel pads as pallet nibs. It constitutes the transition to the anchor escapement.

(d) *Detached escapements*

Their characteristic feature is that the reciprocating motion of the governing element (balance) only brings this into contact with the pallets during a fraction of the stroke (symmetrical about the central position); its motion is otherwise unrestrained. In this way, the drawbacks of the recoil escapement (friction losses and wear) are avoided. The action of the detached escapement resembles that of the dead-beat type in that the motive impulses supplied by the 'scape wheel are only transmitted as the anchor passes the central position. The common features of the mechanisms that ensure short-lived contact (only 12°) between balance and anchor are: a forked lever, integral with the anchor, and a pin-shaped member moving with the balance. Whenever this roller mechanism passes its centre, the pin is in contact with the fork.

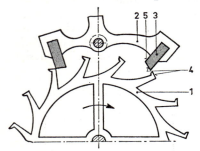

Fig. 1.603. Escapement   with   Swiss
anchor.
1 = 'scape wheel.   4 = impulse face.
2 = anchor.   5 = locking face.
3 = pallet.

For the remainder, the roller mechanisms are designed to provide the following restraints:
- The fork is checked at the extremes of its travel.
- The balance is checked at the extremes of its travel.
- When the anchor, or pallet, is disengaged from the balance, this cannot move out of the position it has assumed in relation to the 'scape wheel.

Several types of detached escapement meeting these requirements are currently available, depending on quality and price (Figs. 1.603 to 1.606 inclusive).

The escapement with *Swiss anchor* (Fig. 1.603), fitted with pallet stones, can be regulated accurately and is therefore used in quality watches (see also Volume 3, Chapter 4). In order to limit the impulse of the anchor as much as possible, the impulse faces are distributed between pallet and 'scape wheel tooth in such a way that the impulse is first received by the pallet, whilst the tip of the tooth slides along the impulse face of the pallet. After this, the impulse or lift is transferred to the tooth whilst its impulse face slides along the pallet. To produce this sequence, the members are so shaped that, at the end of the lift on the exit pallet, both impulse faces are in the same plane.

Although anchors with pallets of equal length or circular pallets, as they are called, were used at one time because they are fairly easy to balance, pallets of unequal length (Fig. 1.604), with which there is less difference in frictional resistance between entrance and exit, are customary nowadays.

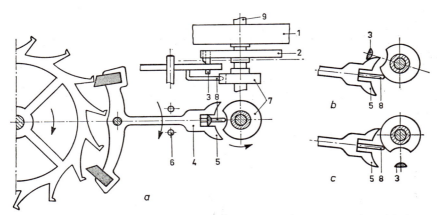

Fig. 1.604. Roller mechanism between Swiss anchor (circular pallets) and balance.

a. Central position.
b. Extreme of balance travel.
c. Locking of fork.

1 = balance.
2 = table roller.
3 = impulse stone or pin.
4 = fork.

5 = horn of fork.
6 = banking pin for fork.
7 = (guard) roller.
8 = (guard) pin (or dart).
9 = balance staff.

Fig. 1.604 shows the roller mechanism between fork and balance associated with a Swiss anchor. The impulse pin (3), or table stone, as it may be called in the case of this stone version, enters the notch between the two horns (5) on the fork (4). The impulse pin is attached to a disc (2) on the balance staff (table roller) and transmits the reciprocating motion of the fork, through about 10°, to the balance shaft. Every time, the pin is thrown out of the fork, enabling the balance to turn through about 420°. The stroke of the anchor is limited by two banking pins (6). The movement of the balance reaches its extreme limit at the position in Fig. 1.604b. The fork is locked in the position of Fig. 1.604c through the shape and setting of the guard roller (7) and the dart (8).

In cheap watches and alarm clocks, an escapement with a *pin-pallet* (Fig. 1.605) is employed. The pallet nibs are round steel pins. Both the locking faces and the impulse faces are on the 'scape wheel tooth. As with the Brocot anchor, the performance is not wholly dead-beat. For this reason, only equidistant pallets are employed.

The roller mechanism between fork and balance associated with the pin-pallet is also shown in Fig. 1.605. The balance pin (8) and the fork (6) engage in the manner already described. The locking of the fork (Fig. 1.605b) takes place as a result of suitable shaping of the horn (7) and the balance staff (9). In the other direction the pin pallets have their stop on the bottom of the tooth space of the 'scape wheel.

The extreme position of the balance (Fig. 1.605c) is governed by the balance pin as it stops against the horn of the fork. The roller mechanism is so shaped that after the departure of the balance pin (Fig. 1.605a) the fork can move somewhat more by the lift of the pallet pins at the 'scape wheel teeth, which is necessary to make this small and simple escapement function properly. This structure is less easy in regulating and therefore only applied in cheap clock mechanisms.

To provide for more accurate regulation, the *Roskopf fork engagement* is employed (Fig. 1.606). The fork locking system is similar to that of the Swiss anchor, but the table roller and stone are replaced by a very much cheaper steel finger.

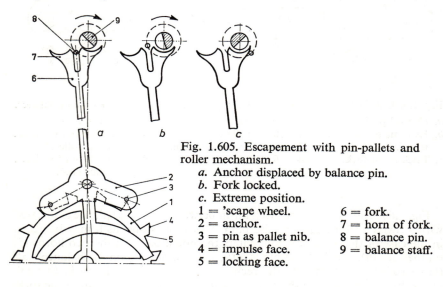

Fig. 1.605. Escapement with pin-pallets and roller mechanism.
a. Anchor displaced by balance pin.
b. Fork locked.
c. Extreme position.
1 = 'scape wheel.   6 = fork.
2 = anchor.   7 = horn of fork.
3 = pin as pallet nib.   8 = balance pin.
4 = impulse face.   9 = balance staff.
5 = locking face.

Fig. 1.606. Roskopf fork engage-ment.
a. Locking of fork.
b. Fork is turned.
c. Extreme position.
    1 = fork.
    2 = horn of fork.
    3 = steel finger.
    4 = (guard) roller.
    5 = (guard) dart.
    6 = balance staff.

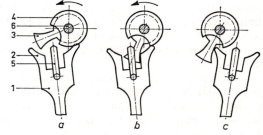

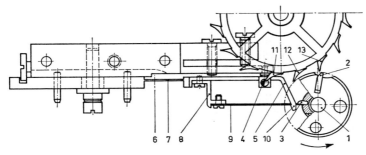

Fig. 1.607. Chronometer escapement.

1 = balance staff.	6 = centre of rotation.	11⎫
2 = impulse (jewel) pallet.	7 = spring hinge.	12⎬ = 'scape wheel
3 = discharging (jewel) pallet.	8 = bracket.	13⎭ teeth.
4 = locking stone (D-shaped).	9 = trip spring.	
5 = lever.	10 = horn on lever (5).	

(e) *Escapement with (flat) spring as anchor*[13]

Very accurate timepieces (chronometers) are used in navigation. They have an escapement of different design, in which the anchor recoils through a flat spring, and where contact with the balance is made as brief as possible. This gives the balance unusual freedom of movement (highly detached), so the instrument can be adjusted very accurately indeed. The chronometer is sensitive to shock: in order to run properly, it must be kept permanently in the same position (universal suspension).

One of the systems employed is illustrated in Fig. 1.607. On the balance staff (1) there are two discs, or rollers, one carrying the impulse pallet (2) and the other the discharging (or unlocking) pallet (3). The D-shaped locking stone (4) is attached to a lever (5) which can rotate about the centre (6) of the spring hinge (7). To the bracket (8) on the lever is attached a very fine, flat spring of gold (9), the trip spring, which rests on the tip (or horn) (10) of the lever and projects slightly.

When the balance staff turns in an anticlockwise direction, the trip spring is moved by the discharging pallet, causing the lever to move by way of the horn so that the locking stone releases the 'scape wheel tooth (11), which it has so far detained. Thereupon, another tooth of the 'scape wheel drops against the impulse pallet, whereby energy is transmitted to the balance. Finally the lever springs back, the locking stone detains the tooth (12) and the discharging pallet turns (now in a clockwise direction) underneath the lever.

The system is constructed to give adjustment of the following:
- The position of the locking stone in relation to the 'scape wheel, both axially and radially.
- The position of the impulse pallet.
- The position of the discharging pallet.
- The position of the tip of the trip spring.

The movements are not checked by any kind of stop.

In instruments for measuring the muzzle (or initial) velocity of projectiles (chronographs), a vibrating flat spring clamped at one end (1) is employed as a self-oscillating anchor (Fig. 1.608). The flat spring engaged by the escape wheel, or siren wheel (2), vibrates at a natural frequency ranging from 200 to 400 c/s, depending on its length and thickness, and passes one tooth during each vibration. Within wide limits, the speed of the instrument is independent of the driving torque applied to the siren wheel. Any abrupt changes in driving torque are compensated by the felt pad (3), pressed home by the spring (4). Very small speed changes can be obtained by varying the pressure of the felt pad: larger ones, by changing the length or thickness of the flat spring. There is scope for readjustment of wear at the tip of the spring.

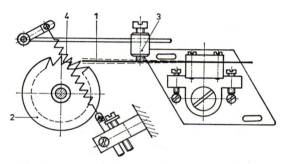

Fig. 1.608. Flat spring as self-oscillating anchor.

1 = flat spring clamped at one end.   3 = felt pad.
2 = 'scape wheel (siren wheel).   4 = tension spring.

Electrically-driven movements in watches (Fig. 1.609) are regulated by a vibrating tuning-fork (1), driven at its natural frequency by current pulses through two coils (2). A magnet (3) and a pole shoe (4) are attached to each of the two prongs of the fork. The coils are in the strong magnetic field set up by the attendant magnet, and containing two windings: a driving winding and a measuring winding. The system has three tasks to perform:
● To convert current pulses into mechanical impulses to sustain the motion of the tuning fork.
● To display (electrically) the amplitude of the tuning fork.
● To determine the time at which the current pulse should pass through the drive winding of the coil.
A stone-tipped driving pawl (5) attached to one of the two vibrating prongs drives a ratchet wheel (6), geared to the hand spindles. With each individual vibration of the tuning-fork prong, this moves the ratchet wheel one tooth forward. A second stone-tipped pawl (8), fastened to the plate (7), ensures that a tooth once passed is checked when the ratchet wheel recoils as a result of friction exerted on it by the returning drive pawl. This system enables the ratchet wheel to be advanced one tooth at a time, despite the fact that the double amplitude of the pawl tip varies between 1 and 3 tooth pitches.

This electromagnetic system includes the following features, to ensure very accurate running:

● The current pulse in the driving coils occurs when the fork prongs have attained their highest velocity, which does not affect the frequency.
● The current pulse becomes stronger as the amplitude declines. Thus, outside influences are quickly eliminated.
● The tuning-fork is made of material having a very low coefficient of expansion. The frequency can be readjusted by turning the two control masses (9).

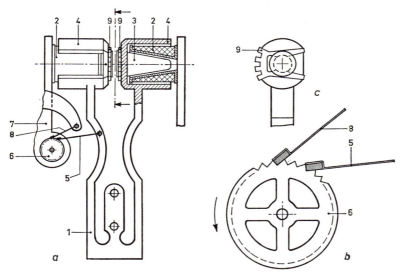

Fig. 1.609. Tuning-fork in regulating system of an electrically-driven watch movement.

    a. Regulating system mounted on plate.   *c*. Enlarged view of control
    b. Ratchet and pawls, enlarged.   mass.

1 = tuning fork.	4 = pole shoes.	7 = plate.
2 = coils.	5 = driving pawl with	8 = fixed pawl with stone tip.
3 = magnets.	stone tip.	9 = rotary control mass.
	6 = ratchet wheel.	

The running accuracy of this movement, unlike that of a mechanically-driven and balance-regulated movement, is unaffected by:

● The position or angle of the instrument.
● The varying driving spring tension.
● The state of lubrication.

A practical design (Accutron) operates at a tuning-fork frequency of 360 c/s and contains a ratchet with 300 teeth.

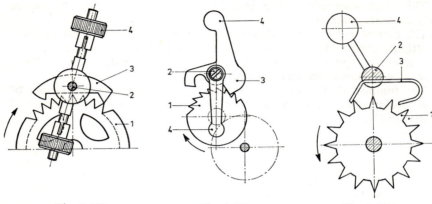

Fig. 1.610   Fig. 1.611   Fig. 1.612

Fig. 1.610. Governor for electromagnetic time relay.
      1 = 'scape wheel.   3 = anchor.
      2 = anchor arbor.   4 = knurled nut as mass.

Fig. 1.611. Governor to determine exposure time.
      1 = 'scape wheel.   3 = anchor.
      2 = anchor arbor.   4 = mass.

Fig. 1.612. Clapper drive of alarm clock.
      1 = 'scape wheel.   3 = nib anchor.
      2 = anchor arbor.   4 = clapper.

(f) *Escapements that are not self-oscillating*

When the speed of the mechanism has to be only approximately constant, the escapement need not be self-oscillating. This means that, instead of travelling from one extreme position to the other of its own accord, the anchor is constrained to do so by the 'scape wheel. The speed depends to some extent on the driving moment, and to a slightly greater extent on the mass moment of inertia of the anchor. Obviously, a recoil escapement is used for this purpose (see also Section (b)). Because the requirements are not very stringent, fairly simple designs are possible. The anchor-bills, which collide time-after-time with the slightly rounded teeth of the 'scape wheel, are of hardened and polished steel.

Escapements of this type are employed in governors for electromagnetic time relays (Fig. 1.610). They include two knurled nuts, for balancing the anchor and controlling the speed slightly (turning both nuts outwards reduces the speed). Similar escapements are used in cameras to govern the exposure time (Fig. 1.611) and in alarm clocks to drive the clapper (Fig. 1.612). In the last-mentioned, the nib anchor can take the form of separate steel strips secured by embedding.

REFERENCES

[1] RICHTER, V. VOSS and KOZER, *Bauelemente der Feinmechanik*, Verlag Technik, Berlin, 1954.

[2] K. HAIN, *Die Feinwerktechnik*, Fachbuchverlag Dr. Pfanneberg, Giessen, 1953.

[3] G. SCHLEE, *Feinmechanische Bauteile*, Verlag Konrad Wittwer, Stuttgart, 1950.

[4] K. RABE, *Grundlagen Feinmechanischer Konstruktionen*, Anton Ziemsen Verlag, Wittenberg/Lutherstadt, 1942.

[5] W. PFERD, *Governors for Dials*, Bell Laboratories Record, 1954, February, p. 69.

[6] H. KALLHARDT and W. UHDEN, *Zur Berechnung und Konstruktion von Fliehkraft-Bremsreglern*, Feinwerktechnik, 1966, 1, p. 18.

[7] H. SCHUSTER, *Fliehkraft-Bremsregler für mechanische Laufwerke*, Feinwerktechnik, 1954, 9, p. 279.

[8] J. H. BICKFORD, *How to Control Speed with Mechanical Governors*, Machine Design, 1967, April 13, p. 168.

[9] H. JENDRITZKI, *Moderne uurwerkreparatie*, Heisterkamp, Amsterdam, 1955.

[10] R. REUTEBUCH, *Der Uhrmacher*, W. Kempter, Ulm, 1951.

[11] A. L. RAWLINGS, *The Science of Clocks and Watches*, Pitman, London, 1948.

[12] J. H. BICKFORD, *Mechanisms for Intermittent Rotary Motion*, Machine Design, 1965, December 23, p. 120.

[13] *Boluva Accutron*, Technisches Handbuch, Bulova Watch Co., Biel, Switzerland.

[14] K. H. BERG, *Konstruktionsbeispiele aus der Feinwerktechnik*, Konstruktion, 1956, 5, p. 186.

[15] G. GLASER, *Mechanische und elektrische Fortschaltvorrichtungen*, Die Uhr, 1968, 9, p. 35; 12, p. 13; 14, p. 11.

[16] S. HILDEBRAND, *Feinmechanische Bauelemente*, 1968, Carl Hanser Verlag, Munich.

Chapter 2

Applied Mechanisms

F. C. W. SLOOFF

2.1 Introduction

A characteristic feature of most precision engineering mechanisms is the accurate dimensioning, to give interchangeability for repair and renewal. Precision-engineered equipment usually has to be transportable and therefore quite light. Also, it often has to operate without maintenance. In many cases the required precision is obtained with components that are themselves coarsely toleranced. The accuracy of the mechanism then depends less on the tolerances. For example, motion can be confined to exact limits by using elastic members and fixed stops. Adjustment is common too, in the sense of permanent changes of shape through resetting during (or after) assembly, in order to meet the functional requirements.

One or two precision engineering products will be analysed at length in this chapter, or described as examples of characteristic mechanisms, but space does not permit every conceivable type of mechanism to be described, of course.

2.2 Intermittent mechanisms (in general)

To obtain a steady picture in film photography and projection, the film strip, after being fed forward, must remain absolutely stationary where the intermittent mechanism has taken it at the end of the shift. Before this happens, however, the pull-down member undergoes a retardation a (cm. s^{-2}). If m is the mass of the strip of film to be shifted, the braking force to slow the film must be at least $m.a$. If the braking force supplied, say, by (intentional) friction in the film channel is less than $m.a$, this puts an overload on the pull-down member and a considerable strain on the film perforations. To reduce this strain, claws having several teeth, to transmit the pull to the film jointly and simultaneously, are used in some intermittent mechanisms.

Because of possible film shrinkage (up to 2%) more than one claw is rarely adopted. In fact, the use of more than one claw in intermittent mechanisms is a precaution, to ensure that the equipment functions properly despite damaged perforations. Even so, there is the drawback that the system does not give correct picture register, since any one of the teeth may be responsible for pull-down at a given time.

It is most important for the pull-down member to engage exactly the same film perforations in the projector as in the camera. Any film perforation imperfections are then compensated automatically in projection. So far, however, the cinematograph equipment industry has not standardized on the number of frames between film gate and claw, or sprocket tooth, although such a step would eliminate the effect of imperfect perforations completely. Nevertheless, much has been done to combat the ill-effects of these imperfections.

2.2.1 Claw mechanisms

These mechanisms have a claw which is periodically engaged with the film during the continuous motion of the intermittent mechanism. In Fig. 2.1,

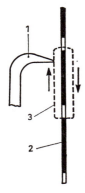

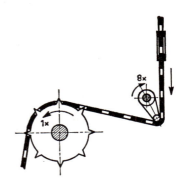

Fig. 2.1. Path of claw tip.
1 = claw tip.
2 = frame of film.
3 = path of claw tip.

Fig. 2.2 Siemens pressure mechanism.

(1) represents the tip of the claw, whose apex describes the path (3) in the direction indicated by the arrows and thus transports the picture length, or frame of film (2). Another type of intermittent movement is the Siemens pressure mechanism of Fig. 2.2, described more fully in Section 2.2.3. The pull-down distance of the film corresponds to the pitch of the film perforations, which is respectively 3·81, 4·27 and 7·62 mm for 8 mm, super 8 and 16 mm equipment. However, the claw describes a path which is longer in the film feed direction than this distance. This is necessary, because there is no way of obtaining a shift exactly equivalent to the pitch, and because there is no other way of bringing the claw safely into engagement with the perforation during its traverse at right-angles to the film feed direction. To limit perforation wear, this traverse should be no longer than is strictly necessary: it should preferably be completed without the claw coming into contact with the film.

As soon as the pull-down is ended, the claw must quickly leave the film. On the other hand, it would be wrong to keep the claw traverse normal to the feed direction down to an absolute minimum, since the part of the claw surface coming into contact with the film would then be the same every time, thus wearing a groove in the claw. Usually, more time is allowed for the retreat of the claw than for its movement in the feed direction. This is done to limit the percentage of time devoted to pull-down.

If the period of a feed cycle is a seconds and the period of actual pull-down is b seconds, the ratio $S = b/a$ may be called the shift ratio. The aim is to procure a low value of S, in order to add to the overall projection time and thus increase the efficiency of the light source.

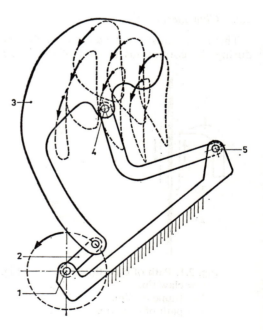

Fig. 2.3. Four-bar linkage as claw mechanism.
1 = centre of crank shaft.
2 = driving crank.
3 = moving system.
4 = tip of pendulum.
5 = centre of pendulum arbor.

Claw mechanisms producing the complete claw path in a very simple way include the four-bar linkage shown in Fig. 2.3. The moving system (3) is driven by a crank (2) pivoting at the fixed point (1) and causing the point (4) to describe an arc of a circle about the fixed point (5). The paths of a number of points in the system (3) are indicated, and it will be seen that they can serve as the path of a possible claw tip. The claw mechanism of the Kodak 16 camera is a practical example of this (Fig. 2.4).

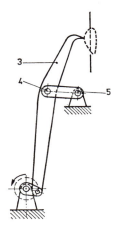

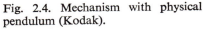

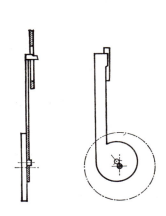

Fig. 2.4. Mechanism with physical pendulum (Kodak).
   3 = claw.
   4 = tip of pendulum.
   5 = centre of pendulum arbor.

Fig. 2.5. Guide in straight slot (Agfa).

A popular method of obtaining the path of point (4) is that involving either a pin guided along a path formed by a slot, as in the Agfa-Movex S and SV cameras (Fig. 2.5) and in the Siemens 16 mm camera (Fig. 2.6), or a member (3) containing a slot which fits round a fixed pin, as in the Austria-Ditmar 9·5 mm camera (Fig. 2.7) and the Eumig C3 camera (Fig. 2.8)*.

If the slot guiding the pin is circular in shape, the mechanism is equivalent to one in which the pin constitutes the pivot of a link, of length corresponding to the radius of curvature of the slot, pin-jointed to a fixed point located at the centre of curvature of the slot. This equivalent (but superior) structure cannot be realized, however, because it is inconsistent with conditions imposed by the film and the film-gate. When a specific claw path is required, the form of the slot can be adapted by resorting to an irregular curvature, but this should be avoided, in view of the tooling problems involved.

To limit the proportional pull-down time, and therefore also the shift ratio (S), a certain amount of play a is sometimes deliberately introduced into claw mechanisms, as shown diagrammatically, to an exaggerated degree,

---

* Note that a transposition of the path point of member (3) and the associated centre of curvature (5) is to be found here (see Volume 1, Section 3.7).

in Figs. 2.9 and 2.10. The drawback of the mechanism of Fig. 2.10 is that
the position of the claw is no longer determinate for every angle of the crank.
Moreover, this system gives rise to infinitely fast accelerations, which have
to be damped largely by the oil in the pin joints. The mechanism in Fig. 2.11
is better as a means of obtaining a low shift ratio. The film is pulled down
only at each downward stroke of the claw (right): this doubles the speed
of pull-down.

In the intermittent mechanisms discussed so far, the motion of the driving
shaft is uniform. By introducing a special intermediate drive, however,
the rotation of the primary shaft can be rendered periodically irregular,
in order to ensure a low shift ratio. This intermediate drive can consist of
a pair of elliptical gears with axes of rotation at the foci.

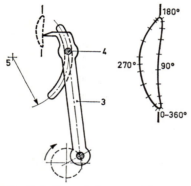

Fig. 2.6. Curved slot as guide
(Siemens).
3 = claw.
4 = pin in claw.
5 = centre of rotation of 4.

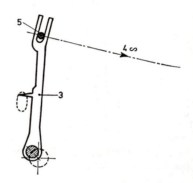

Fig. 2.7. Slot guide round fixed pin
(Austria-Ditmar).
3 = claw.
4 = tip of 3 at infinity.
5 = fixed pin, centre of rotation of 4.

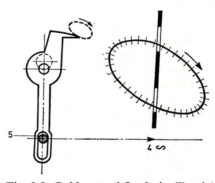

Fig. 2.8. Guide round fixed pin (Eumig)
4 = claw tip at infinity.
5 = fixed pin, centre of rotation of 4.

The anti-parallel four-bar linkage, discussed in Volume 1, Section 3.25, is the equivalent of these elliptical gears with which the reader will doubtless be familiar. Another possible intermediate drive is the pin and fork driver (see Volume 1, Section 2.7.2). A skeleton drawing of this appears in Fig. 2.12.

Finally, the simple four-bar chain with links of equal length pin-jointed at fixed points can also be used as an intermediate drive. AA_0 is a crank rotating at uniform speed. The resultant motion of BB_0 is periodically

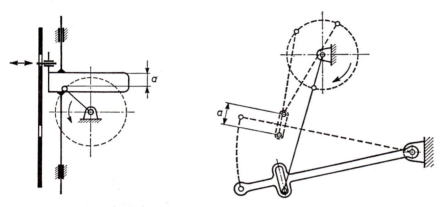

Fig. 2.9. Intentional play in claw mechanism.

Fig. 2.10. Intentional play in claw mechanism.

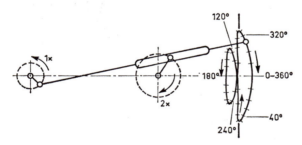

Fig. 2.11. Construction for low shift ratio.

irregular. Different derived positions are indicated in Fig. 2.13 whilst the time scale in the paths of A and B shows how much the speed of B varies. The speed reaches its peak where the spaces between the time divisions are widest. A practical example is shown in skeleton form in Fig. 2.14. The equal-arm four-bar chain is usually employed as an intermediate drive for a Maltese cross (see Section 2.2.4).

2.2.2 The cam as a driver for claw mechanisms

The eccentric cam is much used to drive claw mechanisms, and Fig. 2.15 shows how it can be laid out. The eccentric cam consists of two concentric

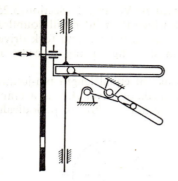

Fig. 2.12. Non-linear intermediate drive.

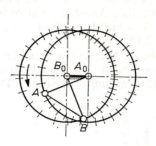

Fig. 2.13. Four-bar linkage as intermediate drive.

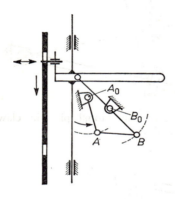

Fig. 2.14. Practical example of four-bar linkage as intermediate drive.

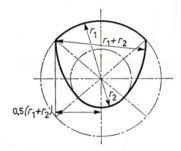

Fig. 2.15. Cam.

arcs of circles with radii r_1 and r_2, linked by two other arcs of circles with radii r_1 and r_2. The periphery of the eccentric cam can be enclosed between two parallel lines at a distance $d = r_1 + r_2$, regardless of the angle of rotation of the cam. Thus, the eccentric cam can be rotated between two parallel guides without clearance and without sticking.

The stroke h is governed by the difference in the radii $(r_1 - r_2)$. The smaller this difference, the "rounder" the shape of the cam. The number of different eccentric shapes that can be used to produce the same stroke h is infinite. For example, Fig. 2.16 illustrates the layout of four different eccentric shapes for the same stroke h. The smallest of them is the Reuleaux triangle of arcs.

It is found that, when an eccentric cam is placed in a square frame restricted to translation, as in Fig. 2.17, this translation follows the paths, shown in Fig. 2.18, dictated by the shape of the cam. The "rounder" the cam, the lower the shift ratio S. Eccentrics having a shift ratio smaller than 1 : 4 can rotate through a certain angle without moving the square frame: this is called the dwell. It will be seen from Fig. 2.19 that anti-clockwise rotation of the eccentric through angle α is concomitant with dwell of the frame.

A neat example of the eccentric cam is to be found in the Zeiss-Ikon-Movikon 16 camera, as shown in Fig. 2.20. The claw tip describes a curved rectangle. The path of the claw tip can be brought nearer to the ideal rectangular shape by increasing the distance a from the fixed pin in the slot to the pivot of the eccentric cam, as indicated in Fig. 2.21.

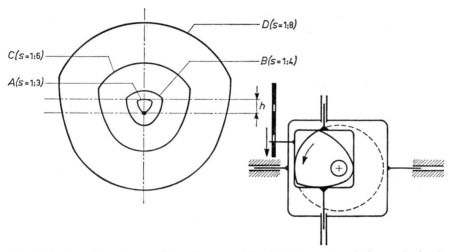

Fig. 2.16. Eccentric shapes giving the same stroke h.

Fig. 2.17. Two translations obtained with a single eccentric cam.

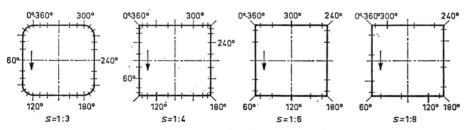

Fig. 2.18. Claw paths derived from eccentric shapes.

The shift ratio can be deduced from Fig. 2.20 in the form of a graph.

Fig. 2.22 demonstrates another method of using the eccentric cam in intermittent mechanisms.

A drive combining an eccentric cam and an eccentric disc, whereby the movement of the claw tip takes place in a plane at right-angles to the driving spindle, is to be found in the Maurer 16 mm and Klangfilm Minikord-V 16 cameras (Fig. 2.23). The eccentric disc, which in Fig. 2.23 drives the claw

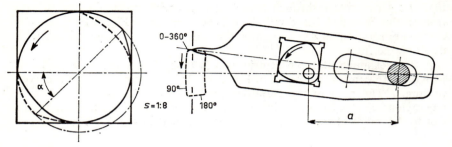

Fig. 2.19. Dwell of eccentric cam frame during rotation through angle α.

Fig. 2.20. Neat example of the eccentric cam (Zeiss-Ikon).

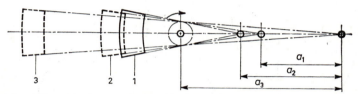

Fig. 2.21. Effect of distance from fixed pin to centre of rotation on the shape of the claw path.

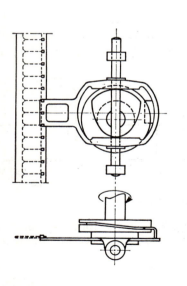

Fig. 2.22. Eccentric combined with indexing cam.

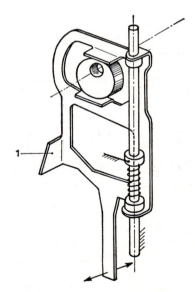

Fig. 2.24. Claw drive (Agfa)
1 = claw.

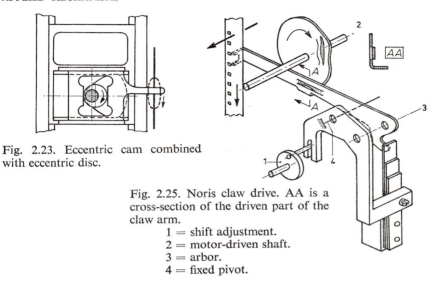

Fig. 2.23. Eccentric cam combined
with eccentric disc.

Fig. 2.25. Noris claw drive. AA is a
cross-section of the driven part of the
claw arm.
  1 = shift adjustment.
  2 = motor-driven shaft.
  3 = arbor.
  4 = fixed pivot.

at right-angles to the feed direction, is often used also to supply the drive
in the feed direction, although this results in a shift ratio of 0·5. This constitutes what is probably the simplest intermittent drive construction to manufacture. In any case, it is the shutter, not the shift ratio, that governs the
exposure time in cameras.

Fig. 2.24 shows the claw drive of the Agfa Movex 8 mm camera, 1 being
the claw tooth. An identical system is found in the Agfa Movexoom camera.
A low shift ratio is more important in projectors. In non-professional
equipment, the movements of the claw in, and at right-angles to, the feed
direction are derived from a single cam, which has a radial, as well as an
axial, projection for the purpose. Fig. 2.25 illustrates the practical application
of this in the Noris 8 mm projector. The solitary claw pivot means that the
motion of the claw prongs is not linear, which is immaterial during the dark
period and gives scope for very simple shift adjustment. Cams have been
exploited in sub-standard film projectors, to design mechanisms which
can be described categorically as precision engineered and combining
optimum performance with minimum production costs. Equipment of a
more professional standard contains mechanisms that are more refined
technically, but also more expensive. An example is the Siemens 2000
projector for 16 mm film. Fig. 2.26 is a skeleton drawing of the twin-cam
drive of the three-pronged claw and Fig. 2.27 illustrates the effect of this.
The shift ratio is 1 : 8. At 24 frames per second, the time required for
pull-down, from 3 to 4, is only 5·4 ms.

2.2.3 Pressure mechanisms

Some projectors have a mechanism that *presses* the film forward intermittently, one frame at a time: the pressure of a loop in the film shifts this

forward one frame at a time in the film channel, as shown diagrammatically in Fig. 2.28. The film (1), drawn forward continuously by the sprocket (2), is propelled along the film channel frame-by-frame by the dog or pusher (6). During part of the feed cycle, there is no driving contact between the dog (6) and the film, so this forms a loop hanging between the film channel and the sprocket. At a given moment, the dog descends on to the loop, shortened before this time, thereby initiating the film shift. This involves a certain amount of sliding between film and dog (6), and the latter is therefore shaped to make contact only with the edges of the film. For this reason, the mechanism became obsolete with the introduction of the sound track in the film margin.

Fig. 2.26. Skeleton drawing of three-pronged claw drive (Siemens).

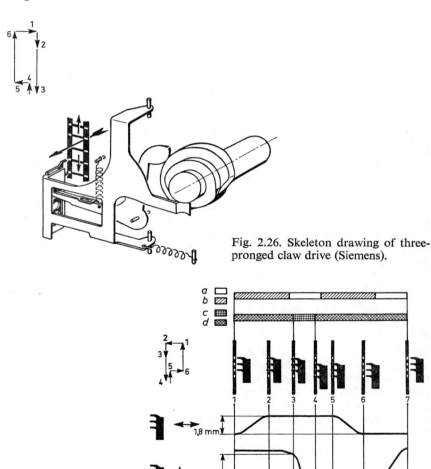

Fig. 2.27. Horizontal and vertical shift motions of three-pronged claw (Siemens).

To obtain a low shift ratio with the pressure mechanism, Siemens have evolved a special version which includes a claw. The claw tip (12a) engages the film during a small part of the cycle of lever (8) (Figs. 2.29 and 2.30). As the film advances, a loop is formed between claw tip and sprocket. The motion of lever (8) stems from the spindles (1) and (2), the latter rotating at a speed which is a whole multiple of the angular velocity of spindle (1). The pressure mechanism with a claw does not, of course, permit any sliding, since the claw fixes the position of the film.

The Siemens mechanism shown in Fig. 2.30 has a shift ratio $S = 1 : 8$, whilst that for the mechanism in Fig. 2.29 is $S = 1 : 6$.

2.2.4 Geneva movements

Geneva movements engage the film continuously, and feed it forward by their own intermittent motion. This motion is derived from the uniform motion of a spindle inside the camera or projector through the action of special members. The member generally used for this purpose is the Maltese

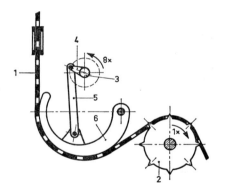

Fig. 2.28. Diagram of a pressure mechanism.
1 = film.
2 = sprocket on driving spindle.
3 = fast driving spindle.
4 = crank.
5 = lever.
6 = dog.

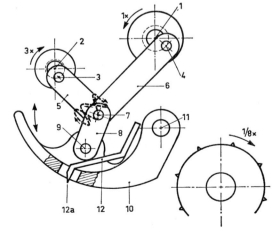

Fig. 2.29. Pressure mechanism with claw (Siemens).
   1 = driving spindle.
   2 = fast driving spindle.
   3, 4 = crank pins.
   5, 6 = links.
   7 = fulcrum.
   8 = lever.
   9 = centre of rotation.
   10 = dog.
   11 = fixed pivot.
   12 = claw.
   12a = claw tip.

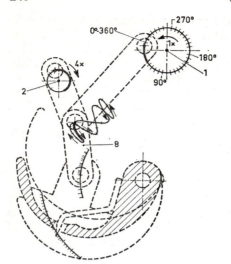

Fig. 2.30. Pattern of motion of pressure mechanisms with claw (Siemens).

cross (Fig. 2.31), the driving element of which rotates at uniform speed on the spindle (1), carrying a pin (3) that enters one of the slots cut in the cross. When the cross has four quadrants, as in Fig. 2.31, each revolution of spindle (1) rotates spindle (2) through 90°, with a shift ratio $S = 1 : 4$.

The mechanism must be so constructed that pin (3) enters the slot radially On completion of the shift, the cross must at once become absolutely stationary. This can be achieved by making the long circular arc of segment (4) on spindle (1) lock perfectly against the cross.

The relationship between the shift ratio S and the number of slots (z) in the cross is

$$S = \frac{z - 2}{2z} \quad \text{or} \quad z = \frac{2}{1 - 2S}.$$

With spindle (1) revolving at 24 revolutions per second and a film shift of 7·62 mm at a time, the variations of the path s, velocity v and acceleration a of the film engaged with the periphery of the sprocket on spindle (2) (Fig. 2.31) make the pattern shown in Fig. 2.32.

To reduce the shift-ratio S of the Maltese cross mechanism, it is sometimes preceded by a pin-and-fork driver (see Volume 1, Section 2.7), as shown in Fig. 2.33. A four-bar chain can also be used as an intermediary between driving spindle and cross, as in Fig. 2.34. The intermittent member here is no longer the original Maltese cross. The path of the pin cuts radially across the circumference of the cross. In this case, the system that locks the cross is more intricate: it has been omitted from Fig. 2.34 to avoid complication.

A method of obtaining intermittent rotation, well known in mechanical engineering, is by means of an indexing mechanism. The uniform-speed shaft of an indexing mechanism usually crosses the intermittently-rotating shaft at right-angles. Fig. 2.35 shows the indexing mechanism of the Lytax projector for 16 mm film. The driving spindle (1) carries the indexing wheel (2), to which a curved tooth (4) is attached. One segment of the inter-mittently-rotating slotted disc (5) fits into the arc between the start and end of the curvature of (4), covering 60° of the circumference of (2). The shift

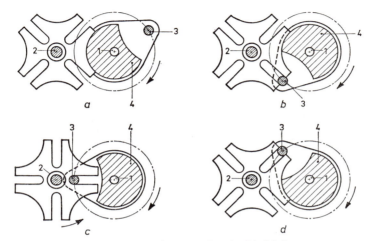

Fig. 2.31. Pattern of motion associated with Maltese cross.

1 = driving spindle.   2 = follower spindle.   3 = pin.   4 = segment.

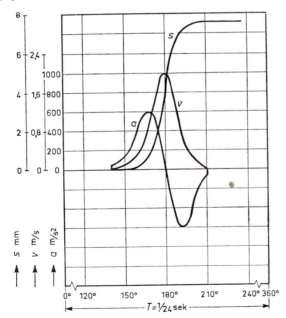

Fig. 2.32. Maltese cross with radial entry.

s = film shift.   v = shift velocity.   a = shift acceleration.

ratio is therefore 1 : 6. The quality of Geneva and indexing mechanisms is governed by the accuracy with which the intermittently-rotating member is divided into sectors and by the solid blocking of this member at rest In the Geneva and indexing mechanisms discussed here, the accuracy of the sector layout depends on the method of distribution, or spacing, employed.

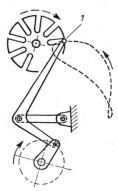

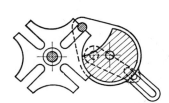

Fig. 2.33. Maltese cross with pin-and-fork as intermediate drive.

Fig. 2.34. Maltese cross with four-bar chain as intermediate drive.

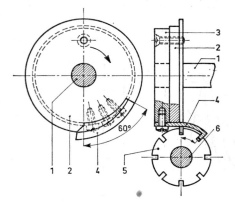

Fig. 2.35. Indexing mechanism (Lytax).
1 = driving spindle.
2 = indexing wheel.
3 = shift dog.
4 = curved tooth.
5 = slotted disc.
6 = film feed spindle.

the solid blocking at rest depends on a minimum of play between the members. Extremely accurate division of the intermittently-rotating member is found in the Philips 16 mm projector Type EL 5000. As Fig. 2.36 shows, this is accomplished by clamping accurately-ground pins between two rings. The pins are hollow and allow elastic compression during assembly: there are 24 pins in all, and alternate pins project. The 12 projecting pins can engage the nylon indexing cam having a shift ratio 1 : 6. The grooves in the indexing cam are slightly narrower than the diameter of the pins, which introduces an element of interference between them and thus ensures very solid blocking. Accurate layout of the intermittently-rotating member, usually coupled direct to the sprocket, is only effective if this member and the sprocket are both free from wobble. In this Philips projector, sprocket eccentricity is therefore limited to a maximum of 2 μm.

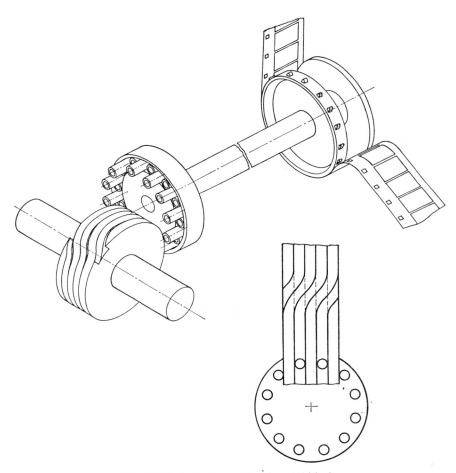

Fig. 2.36. Indexing mechanism (Philips).

REFERENCES

[1] H. WEISE, Kinogerätetechnik, Leipzig, 1950.
[2] J. J. KOTTE, *A professional cine projector for 16 mm film*, Philips Technical Review, 16, 5/6, p. 155.
[3] H. MASCHGAN, *Der neue 16 mm Schmalfilm Projektor "2000"*, Siemens Zeitschrift, 38 (1964), 4, p. 318.
[4] K. ENZ, *Filmprojektoren, Filmprojektion*, Fotokinoverlag, Halle, 1965.
[5] G. PIERSCHEL, *Neue Projektoren für 8 mm Schmalfilm aus dem VEB Kamera- und Kinowerke Dresden*, Feingerätetechnik, 11 (1962) H3, p. 105.

2.3 Governors

Governors are used in sound equipment, film cameras, telephones and so on to keep the angular velocity of the critical shaft within reasonable limits (see also Section 2.1). In the case of electrically-driven shafts, the governor may control the motor supply. Usually, however, the action of the governor is based on strict control of the torque (see Volume 1, Section 2.3.14).

As well as keeping the speed of rotation fairly constant, the governor must often allow for its adjustment too. In film cameras, the frame frequency can be varied from 8 to 64 c/s, without changing the transmission ratio from clockwork or motor to feed mechanism.

An additional factor is the direct relationship between frame frequency and exposure time.

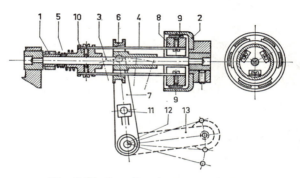

Fig. 2.37. Centrifugal governor (Bauer).

1 = pinion.	6 = adjusting ring.	11 = screw.
2 = brake drum.	7 = lever.	12 = spindle.
3 = governor spindle.	8 = weights (centrifugal mass).	13 = lever.
4 = flat spring.	9 = brake lining.	
5 = torsion spring.	10 = spring holder.	

2.3.1 Governors in cinematograph equipment.

Figure 2.37 shows a governor used in film cameras. The effective length of the flat springs can be varied by lever (7), which affects the spring constant (see Volume 1, Section 2.2.6) and therefore also the centripetal force. The governor is coupled to the camera drive via the pinion (1). In stopping this drive, it is necessary to avoid any jolt, caused by the kinetic force (see Volume 1, Section 2.4.8) of the governor weights (8), that might damage the mechanism. After the drive stops, and therefore also pinion (1), the governor can run down gradually, thanks to the spring (5) between pinion and spring holder (10), which functions as a one-way coupling. As the governor continues to rotate, the spring is loaded contrary to its winding direction, which reduces the frictional torque between spring (5) and pinion (1) appreciably, and enables the governor to run down (see also Volume 1, Section 2.3.7).

The characteristic feature of the governor in Fig. 2.37 is that the outward swing of the weights (8) is kept within constant limits by the wall of the fixed part of the brake drum (2).

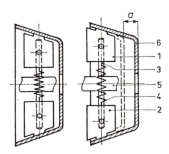

Fig. 2.38. Centrifugal governor (principle).
1, 2 = centrifugal mass.
3 = guide pin.
4 = springs.
5 = spindle.
6 = brake drum.

In contrast, the continuously-variable governor can be constructed on a principle that does not involve constant limitation of this swing. The result is a conical governor, shown in two different states, corresponding to different speeds, in Fig. 2.38. The actual adjustment is made by shifting the solid part of the brake drum (6) axially a distance a. Figure 2.39 illustrates a practical example. The turret (20) can be turned to different settings by the adjusting knob (22). By means of the adjusting screws (19) in the turret (20), the guide (10) of the brake cone (2) of the governor is brought to the correct axial position, by moving it along the guide pins (12). Some particulars of the self-braking action of this type of guide have already been given in Volume 1, Section 2.3.8. The speed setting of the governor can be changed without bringing the centrifugal weights (4) into contact with the brake cone (2).

Fig. 2.39. Centrifugal governor
(Emel).
 1 = pinion.
 2 = conical brake drum.
 3 = bush.
 4 = weights (centrifugal mass).
 5 = brake lining.
 6 = hinge pins
 7, 8 = flat springs.
 9 = pin.
10 = guide.
11 = spring.
12, 13 = guide pins.
14 = fixed shaft.
15 = frame.
16 = lever.
17 to 19 = adjusting screws.
20 = turret.
21 = dial disc.
22 = adjusting knob.
23 = spring.
24 = wall of apparatus.

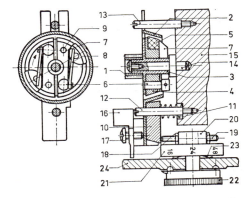

2.3.2 Governors in telephones

Telephone dials supplied in bulk by German companies have a governor operating on the principle discussed in Section 2.3.1.

The Bell Telephone Company have developed a new dial governor for subscribers' instruments, based on the principle discussed in Volume 1, Section 2.3.14, but with a very much stiffer action, obtained through a special method of torque transmission and through rotation in the opposite direction from Fig. 2.96 (Volume 1), namely clockwise. Figure 2.40 illustrates the mechanism, in which shaft (1) is driven clockwise. The drive bars (2) are attached to this shaft and their tips press against nibs (3) on the centrifugal weights (4). A friction stud (5) on these weights runs against the inside of the fixed friction drum (6). Spring (7) supplies a force tending to drive the weights (4) inwards.

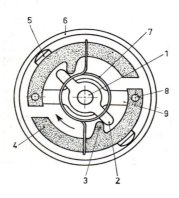

Fig. 2.40. Centrifugal governor (Bell Telephone Company).

1 = shaft to be governed.
2 = drive bars.
3 = nibs on 4.
4 = centrifugal weights.
5 = friction stud on 4.
6 = brake drum.
7 = restraining spring.
8 = pivot pin.
9 = fly bar.

The pivot pins (8) of these weights are attached to a member (9) which rotates freely on the shaft (fly bar). Figure 2.41 shows the pattern of forces associated with the new governor. The slight restraining force of spring (7) is ignored. It will be seen that, in the old governor, the object of forces R and W was to produce rotation of the weight about pivot pin (8), contrary to the effect produced by the centrifugal force C. In the new governor (Fig. 2.41b), R, W and C all contribute to rotation in the same direction about the pivot (8). The result is that, with ten times more external (input) torque on shaft (1), the speed is only 25% above that for normal torque. With the old governor, the difference in speed was 81%.[1]

2.3.3 Electro-mechanical governors

There are a number of neat solutions to the problem of stabilizing the angular velocity of a motor shaft, a common feature of which is that the voltage from a generator coupled to the motor to be governed is compared with a constant reference voltage. The difference voltage thus detected governs the motor supply, which is corrected to bring the difference voltage to zero (see Chapter 4 and Volume 7, Chapter 5).

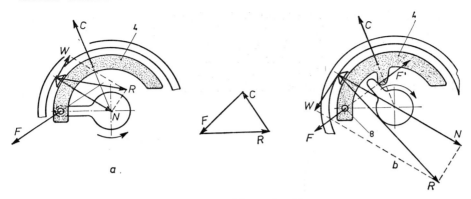

Fig. 2.41. *a* = pattern of forces in old governors.
b = pattern of forces in Bell governors.

Electro-mechanical speed control is used in electrically-driven instruments whose speed of rotation has to be kept within certain limits as, for instance, in teleprinters. A type of centrifugal governor is adopted for this purpose (Fig. 2.42). Slip rings and wipers (4) and (5) give electrical access to a system of contacts (1), (2) and (3) mounted on the motor shaft. At low speed, the series resistor (6) is short-circuited by contact (1): when the speed builds-up unduly, contact (1) opens, bringing series resistor (6) into the armature circuit. Resistor (2) and capacitor (3) serve to suppress sparking at the contact (1) (see Volume 1, Chapter 5.11). Figure 2.43 shows how the contact (1) functions at the correct speed of rotation. This contact is carried on the member (3), rotating about (2), whose contact end (1) is held against the mating contact by an adjustable spring (4). The following forces act on member (3) in Fig. 2.44:

a = the contact force.
b = the tensile force of spring (4).
c = the centrifugal force.
d = the weight, or force of gravity.
e = the bonding force at (2), not indicated in the diagram.

Fig. 2.42. Electromechanical governor (circuit).

1 = switch contacts.
2 = resistor.
3 = capacitor.
4, 5 = slip rings.
6 = series resistance.
7 = field winding.
8 = variable resistor.

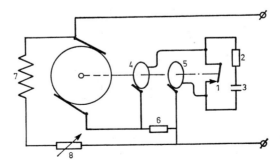

Due to the angular velocity ω, all these forces, except that of gravity d, revolve with member (3). The pattern of forces can therefore be represented more simply by assuming (3) to be stationary and assigning a rotation— to the vector d. This is shown in Fig. 2.45. Determination of the resultant torque of forces b, c and d on member (3) reveals it to be the torque exerted by the resultant f about the centre of rotation (2).

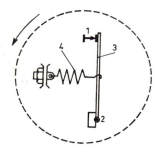

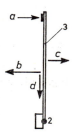

Fig. 2.43. Electro-mechanical governor (mechanism).
1 = contact.   3 = rotating member.
2 = centre of   4 = spring.
    rotation.

Fig. 2.44. Electro-mechanical governor (forces).
2 = centre of rotation
3 = rotating member.
a = the contact force.
b = the tensile force of spring (4).
c = the centrifugal force.
d = the weight, or force of gravity.

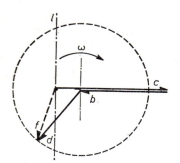

Fig. 2.45. Electro-mechanical governor (polygon of forces).

Since the tip of d is on the circumference of a circle, so also is the extreme of the resultant f. The torque of f must be compensated by that of the contact force a. A positive contact force a occurs only when the vector point of d, and therefore also of f, is on the left of line l. On the right of this line, there is therefore a negative contact force: in other words, the contact opens and a state of equilibrium is reached through a periodic change in the force b. The contact therefore "breaks" once during each revolution, thereby producing an average armature current and keeping the motor at the correct speed. Since any derangement would necessitate adjustment of the mechanism inside the governor, provision is made for electrical regulation by the variable resistor (8) in the field circuit.

REFERENCES

[1] W. PFERD, *Governors for Dials*, Bell Laboratories Record, Feb. 1954, p. 69.

2.4 Operating principle of the sewing machine

Most sewing machines make a lock stitch, as shown in Fig. 2.46a. This takes the least amount of sewing thread, compared with other stitches, but is unsuitable for knitted fabrics and other stretchable materials, because of its lack of elasticity.

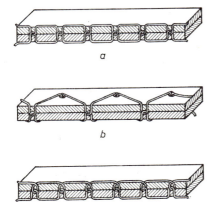

Fig. 2.46.
a. Machined seam-lock stitch.
b. Zig-zag stitch.
c. Back-tack.

However, the zig-zag stitch shown in Fig. 2.46b does possess the necessary elasticity for this purpose. It is perhaps not very widely known that, when operated in reverse, most lock stitch sewing machines produce another kind of "twisted lock stitch", or back-tack as it is called, indicated in Fig. 2.46c. This is due to the position of the loop catcher, or race hook, relative to the needle in its direction of travel. A lock stitch is produced as follows; the needle is thrust into the material to be sewn, carrying the top thread, with the loop at its tip, through the material and down to the lowest position of the needle. When the needle rises again, the top thread on one side (the left in Fig. 2.47) cannot emerge (or, at any rate, cannot

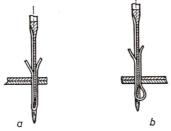

Fig. 2.47. Positions of the sewing needle. *a.* Lowest position. *b.* Formed loop.

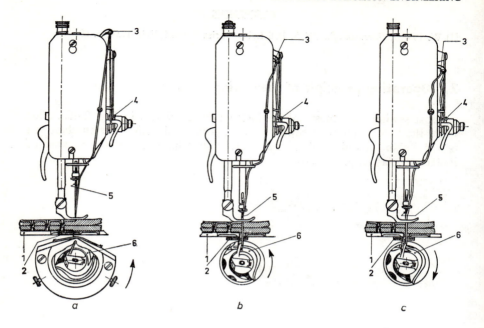

Fig. 2.48. How the lock stitch is formed in a machine with loop hook.

1 = throw plate.   3 = take-up lever.   5 = needle.
2 = feed.   4 = thread tensioning spring.   6 = hook.

a. 1. Needle enters material.
   2. Hook still moving back (anti-clockwise).
   3. Take-up lever begins to descend.
   4. Feed has completed its index and drops below work surface.

b. 1. Needle has reached its lowest position.
   2. Hook about to begin clockwise rotation.
   3. Take-up lever has completed 1/3 of its downstroke, and leaves top thread
      completely relaxed.
   4. Feed has reached its lowest position.

c. 1. Needle has begun its ascent.
   2. Tip of hook opposite centre of needle, about 1·5 mm above top edge of
      needle slot.
   3. Take-up lever moves further down.
   4. Feed is returning under work surface.

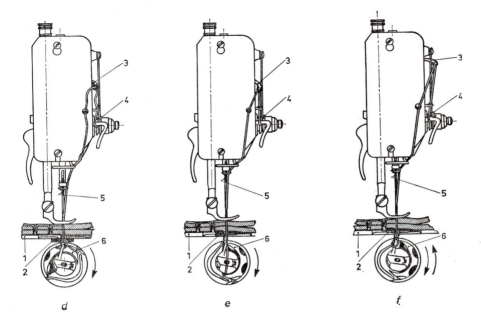

Fig. 2.48 (continued). How the lock stitch is formed in a machine with loop hook.

1 = throw plate.   3 = take up lever.   5 = needle.
2 = feed.   4 = thread tensioning spring. 6 = hook.

d. 1. Needle rises rapidly.
   2. Hook has engaged top thread loop and guides this over case containing bottom thread bobbin.
   3. Take-up lever continues to rise.
   4. Feed, having reached the end of its return, rises and protrudes above the work surface.

e. 1. Needle is near its highest position.
   2. Hook has reached the end of its clockwise rotation.
   3. Take-up lever rises quickly to tighten top thread, as this slips from the hook.
   4. Feed has reached its highest position.

f. 1. Needle has reached its highest position and pauses.
   2. Hook begins reverse rotation.
   3. Take-up lever has completed 2/3 of its upstroke and top thread slips off the bottom thread case.
   4. Feed begins to take the material forward.

emerge as readily) from the material, because it is trapped between this and the needle, and there is more friction between thread and material than between thread and needle.

A longitudinal channel in the other side of the needle accommodates the top thread carried along by the needle. The result is that, in the first phase of needle retraction, the thread forms a loop on the straight side of the needle. The lock stitch is now formed, by passing the bottom thread bobbin right through the loop and then, when the needle has been withdrawn, drawing the top thread tight with sufficient force to drag the bottom thread halfway into the material. Various stages in this process are illustrated by simple sketches in Fig. 2.49a. During each cycle, the tip of the shuttle enters the loop and widens it. The shuttle itself lies in its holder in such a way as to enable the top thread to slip completely round it. There is another method of passing the bottom thread through, whereby a loop catcher hooks the loop, opens it out and wraps it round the bottom thread bobbin. The bobbin takes no part in the motion of the loop hook, or claw (Fig. 2.48).

The hook motion is a rotation with one or more revolutions per cycle of the needle. The hook takes proportionately less time to loop the top thread round the bottom thread bobbin when revolving more than once, e.g. twice

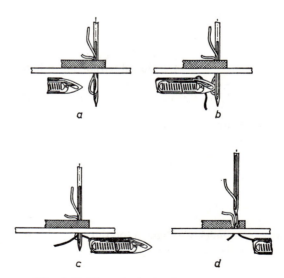

Fig. 2.49. How the lock stitch is formed.

a. As it rises, the needle forms the top thread loop.
b. The loop catcher (in this case the shuttle) slides into the top thread loop.
c. The shuttle has passed through the top thread loop and the loop is tightened.
d. The needle has moved further upwards and the take-up lever draws the bottom thread into the material, along with the top thread.

per needle cycle, than with one revolution per cycle. In shuttle machines, the needle continues to descend a short distance after the loop has been formed, to facilitate opening out. This motion is best imparted by a cam. In all sewing machines, it is most important for the hook path to remain at the correct distance (about 0·1 mm) from the needle, even if this happens to buckle due to rough handling. To keep it so, the needle, in descending, first passes through a guide hole correctly located in relation to the hook path. In principle it would not be possible to register the needle, linked by various intermediate members to the main shaft, or arm, of the machine, exactly opposite a slot in the machine base, without some form of adjustment. In most sewing machines, this adjustment consists of hammering a permanent deformation in the arm. A less rough-and-ready (and technically more acceptable) method is employed in the "Elna", which has an eccentric and a locking screw to provide continuous adjustment. The tension in the top and bottom threads is obtained through friction. Both threads have to pass devices which draw them under spring pressure between two metal surfaces.

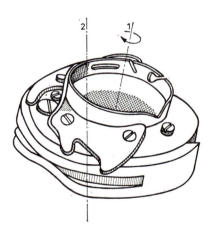

Fig. 2.50. Elna bottom thread bobbin case.
    1 = axis of rotation of hook.
    2 = axis of needle.

In zig-zag machines, the needle moves not only vertically, but also, after each stitch, at right-angles to the sewing direction. Thus, stitches are formed at two adjacent positions on the machine base. Loops are therefore formed in two different places. In zig-zag machines with a loop hook, the hook rotates on a basically vertical axis, enabling it to engage the loop at both positions of stitch insertion. With the hook rotating at uniform speed, it can only assume the correct angle relative to the loop in one of these two positions. Since the loop is formed during the ascent of the needle, the hook of a zig-zag sewing machine either intercepts the right-hand stitch too soon, or the left-hand stitch too late. This problem has been solved very neatly in the "Elna" sewing machine by placing the hook spindle at an angle to the vertical, so that the hook intercepts the right-hand and left-hand loops at the appropriate level for the time of its passage (Fig. 2.50). This will be discussed more fully in Section 2.4.3.

2.4.1 Top thread take-up of the sewing machine

The top thread take-up mechanism, or take-up lever, has the task of shortening or lengthening the span of top thread between the tensioning clamp and the eye of the needle. This lengthening facilitates the widening of the loop by the hook or the shuttle: shortening draws the top thread through the material, bringing the bottom thread into it. The importance of the take-up function has fostered a variety of design solutions, some of which will now be described.

(a) Needle bar as take-up regulator

Fig. 2.51 illustrates the direct use of the needle bar (1) to regulate thread take-up in the original Howe* machine. The top thread passed through an adjustable thread guide (2) and through a hole in the needle bar. The amount the span shortened as the needle rose and could be varied by vertical adjustment of the thread guide.

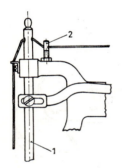
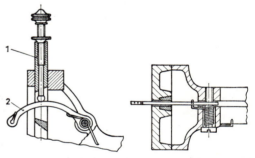

Fig. 2.51. The needle bar as take-up regulator.
1 = needle bar.   2 = thread guide.

Fig. 2.52. Take-up regulator with torsion spring.
 1 = needle bar.
 2 = take-up lever (regulator).

(b) Spring-loaded take-up regulator

Singer introduced an improved thread take-up in the form of a needle-bar regulator, as shown in Fig. 2.52. The needle bar (1) drives the take-up lever (2) down, whilst the upward thrust of this lever is supplied by a spring.

The setting of the take-up lever is adjustable by a screw in the needle bar (1). Spring-loaded take-up levers are often used in shoemakers' machines.

(c) Take-up regulator with cam or indexing cam

None of these solutions was altogether satisfactory, since they did not determine the path of the take-up lever as a function of time.

House was the first to effect an improvement in this respect, by using an indexing cam to control the take-up system, in the manner of Fig. 2.53.

---

* An American, Elias Howe, constructed the first shuttle sewing machine in about 1843

This enabled the phase of the take-up lever (2) and needle bar (1) movements to be varied. Because of the shape of the indexing cam, House was able to establish the position of the take-up lever at every level of the needle bar. Although theoretically ideal, this solution had certain practical drawbacks. The fast accelerations of the mechanism set up inertia forces (see Volume 1, Section 3.27) which caused wear, so the House system is suitable only for slow-running machines, up to about 1 000 stitches per minute (Phoenix, etc.).

(d) *Four-bar chain construction*

In 1892, the quest for higher speed led Singer to mechanisms with lighter members, pin-jointed together on the principle of the four-bar chain (see Volume 1, Section 3.1.4). Figure 2.54 shows a constructive solution to the problem. The main shaft (1) drives two cranks, one of which (2) actuates the needle bar and the other (3) controls the take-up lever, which is also guided by an arm (4). The path of the eye in the take-up lever is indicated in the diagram. This solution to the problem of take-up regulation found the most acceptance. With the take-up lever (5) pivoted in needle bearings, speeds up to 5 000 stitches per minute can be attained (Singer, Dürkopp, Pfaff, Union, etc.).

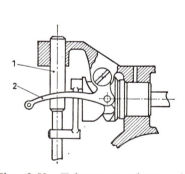

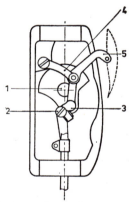

Fig. 2.53. Take-up regulator with indexing cam.

    1 = needle-bar.
    2 = take-up lever (regulator).

Fig. 2.54. Take-up lever as part of four-bar chain.
    1 = main shaft.
    2 = crank actuating needle-bar.
    3 = crank actuating take-up lever.
    4 = guide arm with fixed pivot.
    5 = take-up lever.

(e) *Take-up regulator with slide*

This system was evolved by Wheeler and Wilson and produced in 1901. Figure 2.55 shows that the common crank (2) for needle-bar and take-up lever on the main shaft (1) carries a crosspiece (3) which constitutes a sliding sleeve for the link (4) with the take-up lever, which has a fixed pivot at point (6). Thus, the path of the eye in the take-up lever is circular. This mechanism is found in fast-running, industrial sewing machines.

(f) *Rotary thread take-up regulator*

Willcox and Gibbs Mfg. Co. made a rotary take-up regulator developed from a principle discovered by Gritzner in 1878. This involves the least disturbance of the uniformity of the rotary motion by forces of inertia, since such forces can only arise from the very slight mass of the thread itself (Fig. 2.56). In this system, the nature of the extension and contraction of the thread span is related to the angle between the thread guide (1) and the point of support-cum-clamping point (2), to which is attached the tension spring, to exert a pull on the top thread during rotation, as indicated.

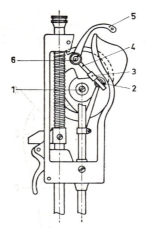

Fig. 2.55. Take-up regulator with slide.

1 = main shaft.
2 = crank.
3 = crosspiece (sliding sleeve for 4).
4 = link of 5.
5 = take-up lever (regulator).
6 = fixed pivot of take-up lever.

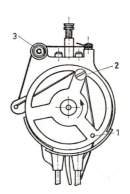

Fig. 2.56. Rotary thread take-up regulator.

1 = thread guide.
2 = support and clamping point.
3 = tension spring to exert a pull on the top thread.

2.4.2 *Feed devices in sewing machines*

In ordinary domestic sewing machines, the work is fed forward by a method Wilson devised in 1852. Considering only the bottom feed, as it is called, we have a toothed member, or feeder, travelling in the manner indicated in Fig. 2.57, namely upwards into the work, horizontally in the feed direction, down out of the work and then horizontally back to the starting position. This motion is obtained by means of eccentrics or cams, or indexing cams, levers, etc.

Fig. 2.57. Feed motion.

1. Upward motion.   3. Downward motion.
2. Feeding work forward.   4. Return.

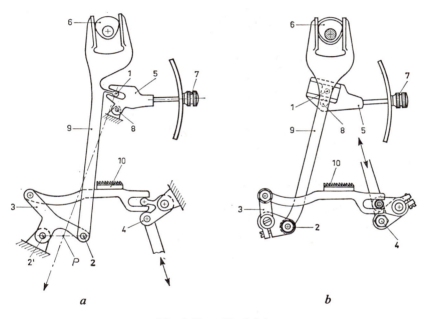

a   b

Fig. 2.58. *a.* Feed drive.

1 = slot guide.   6 = eccentric on top main shaft.
2 = 2, 2′ pin joints.   7 = stitch length adjusting knob.
3 = rocker arm.   8 = pivot of 5.
4 = driven periodically by bottom main   9 = coupling link of four-bar chain
    shaft.   2—2′, 1—.1.
5 = stitch length adjuster.   10 = feeder.

b. Feed drive (practical example).

1 = slot guide.   6 = eccentric on top main shaft.
2 = 2, 2′ pin joints.   7 = stitch length adjusting knob.
3 = rocker arm.   8 = pivot of 5.
4 = driven periodically by bottom main   9 = coupling link of four-bar chain
    shaft.   2—2′, 1—1′.
5 = stitch length adjuster.   10 = feeder.

Provision must also be made for continuous adjustment of the feed stroke per stitch from outside, in order to obtain the desired stitch length. To enable patches to be inserted, the feeder in modern machines must be retractable, to a position where it cannot protrude above the work surface. The feeder owes its grip on the work to the force with which it is pressed against the work surface by the pressure foot. This force is supplied by a spring, which must be adjustable to suit the material being sewn. The feed drive system is illustrated in outline in Fig. 2.58a, whilst Fig. 2.58b gives a practical example. The motion of feeder (10) has a horizontal component, which must be variable for adjusting the stitch length, and a vertical component, which must be derived from the motion of point (4). The latter component is constant, and transmitted by a direct drive via an eccentric on the main shaft. The amplitude and sense of the motion of (3) are controlled by rotating the stitch adjuster (5) about the fixed pivot (8), as follows (see Fig. 2.58a).

Examining member (9), whose fork fits round the eccentric (6), we find that the direction of motion of points (1) and (2) is known, as is the location of the instantaneous centre P of the motion (see Volume 1, Section 2.3.3). This instantaneous centre P is the point of intersection of (2—2') with (1—1'), where (1') is the centre of curvature at infinity of the path of (1). In the situation illustrated, P is between (2) and (2'). When the stitch adjuster (5) is turned in a clockwise direction by the knob (7), P is shifted towards (2'). Coincidence of centre P and (2') means that (3) has no velocity, so it cannot contribute a horizontal component to the velocity of the feeder (10). If P reaches (2—2') produced, that is, on the left of (2'), then (2) will begin to move in the opposite direction under otherwise identical conditions. Thus, the feed will go into its return movement.

Since Fig. 2.58a may be regarded as an intantaneous record, even with (5) in a position at which the centre coincides with (2'), the result of a change in the position of the eccentric (6), not accompanied by a change in the position of (5), will be that the centre P no longer coincides exactly with (2'). Therefore, the position of (5), corresponding to zero feed motion, really provides centre P with a short path extending partly to the left, and partly to the right, of (2'). This corresponds to a shift (during each needle cycle) composed of a short advance and a short return motion, resulting in zero feed.

If point (4) is turned aside in an anti-clockwise direction, the feeder (10) reaches a position in which the serrations are fully retracted below the work surface. This is a situation suitable for patching.

In modern machines, the practice is to accommodate all the feed members and mechanisms in the machine base. Also, every effort is made to reduce precision sliding motions as far as possible to simple rotary guides (pin-joints). For example, Fig. 2.59 shows the feed mechanism of the Elna Supermatic. In the base rotates the horizontal driving shaft (1), which moves the eccentric (2) via a pair of right-angle gears. The eccentric (2) is connected by a driving rod (3) to the intermediate link (4), of which point A is connected by link (5) to point A_0 of crank (6), dowelled to the vertical regulator spindle (7). This spindle (7) is turned to a predetermined position for a given feed rate by means of an external knob, so that crank (6) can be adjusted from outside the machine.

The horizontal component of the feeder motion (10) is derived from the motion of point B of (4), by means of the rocker (8), which pivots about the fixed point B_0. The rocker (8) carries a vertical pin (9), on which the feeder (10) can slide up and down.

The instantaneous centre P of the motion of the intermediate link (4) is identified as the point of intersection of AA_0 and BB_0 (Fig. 2.60). Now the velocity v_B of point B is concomitant with a given velocity v_C imparted

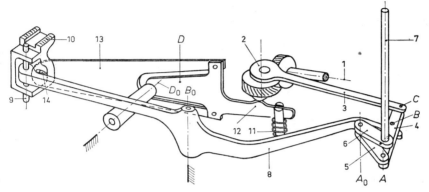

Fig. 2.59. Feed drive of Elna Supermatic.

1 = driving shaft.	9 = slide pin for feeder 10.
2 = eccentric.	10 = feeder.
3 = driving rod.	11 = compression spring.
4, 5 = intermediate link.	12 = follower on curve on eccentric shaft (2).
6 = crank on spindle 7.	13 = flat spring.
7 = stitch-length regulating spindle.	14 = drive pin on 13.
8 = rocker pivoting about B_0.	

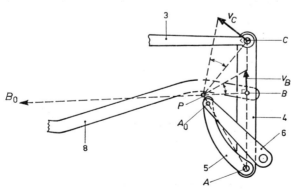

Fig. 2.60. Kinematic analysis of the Elna Supermatic feed drive.

3 = driving rod.	6 = crank on spindle 7.
4, 5 = intermediate links.	8 = rocker pivoting about B_0.

to point C by the eccentric (2). When crank (6) is adjusted to bring A_0 on the other side of (4) then, with C moving in the same direction to the left, point B is constrained to change its direction of motion, thereby shifting the feed into reverse.

No feed will take place when A_0 is brought on to line AB, so that P and B coincide.

With zero feed, the instantaneous centre of the motion of link (4) *invariably* coincides with pivot (B), whatever the position of the eccentric (2), because AA_0 and AB are of equal length. This is the essential difference with the mechanism shown in Fig. 2.58.

The vertical component of the motion of feeder (10) is derived from the cam-shaped bottom edge of the gear wheel carrying the eccentric (2).

Member D, which is free to move about the fixed horizontal axis D_0, is pressed against the cam by the force of spring (11) with the follower (12), and thus undergoes angular rotation about point D_0. To make these angular rotations effective for the vertical movement of feeder (10), which also performs a horizontal movement, member D is extended into a flat spring (13) carrying the driving pin (14). This pin engages an elongated hole in feeder (10) and transmits the vertical motion to it. Member D is held in position horizontally by the guide in a slot in the pin, round which the spring (11) is fitted.

A special feature of the Elna Supermatic is that the feed length, in reverse as well, can be controlled automatically by an exchangeable cam. This, together with the cam-control of the zig-zag deflection, gives scope for decorative stitching characterized by repeated close-stitches and, if necessary, repeated back-stitching. The feed regulating cam then governs the setting of spindle (7).

2.4.3 Zig-zag stitching devices

Consider now those systems in which the material itself makes no sideways movement, and it is consequently the needle that switches periodically from one line of stitch insertion to another. There are two methods of obtaining the zig-zag stitch:

(a) The needle-bar guide oscillates about a fixed pivot in a vertical plane, as in the Phoenix, Anker, Gritzner, Mundlos, Pfaff, Singer, Betz, Haid & Neu, and Husqvarna. Details are given in Fig. 2.61.

(b) The needle-bar guide swings to-and-fro, like a door, on a vertical axis, as on the Adler, Meister, Elna, Anker, Bernina, Husqvarna, Köhler, Necchi, Naumann, Vesta, Weba and Dürkopp (Fig. 2.62).

With method (b), employed by Kayser and Neidlinger as long ago as 1883, thick layers can be sewn without any difficulty, because the up-and-down motion of the needle is not strictly vertical. With method (a), on the other hand, less force is required to accelerate the machine members, since the mass moment of inertia about the pivot axis is smaller.

As mentioned in Section 2.4, the hook axis in zig-zag machines should, in principle, be in the same direction as the vertical motion of the needle.

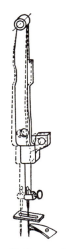

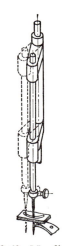

Fig. 2.61. Needle-bar guide pivoting on a horizontal axis.

Fig. 2.62. Needle-bar guide pivoting on a vertical axis.

Since there are two positions where a loop can form during the withdrawal of the needle, it is not possible to bring the tip of a hook pivoted on a vertical axis to the loop at the correct moment in both positions. If the hook reaches the right-hand position at the correct time, it reaches the left-hand position too late: similarly, correct register at the left-hand position means that it reaches the right-hand position too soon. Where the zig-zag is not more than 5 mm wide, a compromise adjustment is therefore accepted, whereby the hook timing is really too soon for the left-hand position, and too late for the right-hand position. This problem has only been solved satisfactorily in the Elna, by slanting the axis of the hook to the right so as to elevate the tip of the hook on its way from the right-hand to the left-hand position, and thus enable it to intercept both the left-hand loop, formed at a higher level, and the right-hand loop at the correct moments (Fig. 2.50).[10]

Perhaps the most valuable feature of a zig-zag machine is its scope for continuous variation of the zig-zag width, not only in relation to the total width, but also in regard to that portion of the zig-zag area where the operator wishes to insert the stitches.

There are various design solutions to this problem, one of which, employed in Adler, Mundlos, Necchi, Phoenix and Gritzner machines, will now be discussed (Fig. 2.63).

The horizontal main shaft in the upper structure of the machine carries a crank-driving rod mechanism to actuate the needle-bar (1) in its guide (2), which is attached to the frame of the machine at pivot (3). Pin-jointed to the guide (2) at (4) is a bar (5) having, at its other end, a pivoted slide which fits in a circular track of member (7). Member (7) is pin-jointed to the frame of the machine at (8), its other end being pin-jointed to member (10). Member (10) is forked to fit round the eccentric cam driven by the main shaft, and slotted to fit round the fixed pin (11).

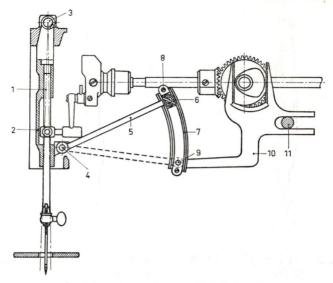

Fig. 2.63. How a zig-zag stitch is formed.

1 = needle-bar.	7 = track for 6.
2 = needle-bar guide.	8 = fixed pivot of 7 on frame.
3 = pivot for 2.	9 = pin-joint between 7 and 10.
4 = pivot for 2.	10 = drive member.
5 = coupling bar.	11 = fixed guide pin.
6 = slide.	

With bar (5) at the top of its stroke, the needle-bar guide is locked, so that no zig-zag stitch is produced. As the slide is moved towards the pin-joint (9) of member (7), progressively increasing deflections are imparted to the needle-bar guide (2).

The variation of the zig-zag shape thus obtained is shown in outline in Fig. 2.64. The motion of member (10) is relatively complex. The instantaneous centre of this motion is the point of intersection of the line through (11) normal to the slot direction there, with the line (8)–(9).

Now, if the fixed pivot (8) is transferred a distance a to the left along an arc of a circle whose radius is the distance from (8) to (9), it will not result in any change in the zig-zag stitch with maximum deflection supplied by point (9), produced a moment ago. Shifting the slide (6) towards the (transferred) point (8) then results in the change of stitch shown in Fig. 2.65. The displacement a of the "neutral" point is caused by the displacement of the fixed pivot (8). Constructively, this displacement a can be made equivalent to half the width of the zig-zag track, which sets a limit to one side of the pattern in Fig. 2.65, parallel to the sewing direction.

In exactly the same way, a displacement of the fixed pivot (8) to the right will produce the zig-zag pattern shown in Fig. 2.66, in response to gradual movement of slide (6) towards point (9).

In the Dürkopp machine, the problem of adjusting the width of the zig-zag pattern and the position of the "zero line" of this pattern has been solved as follows.[2]

Two manual controls are provided at the front of the machine, one to adjust the zig-zag deflection and the other to adjust the neutral setting of the needle-bar guide. The special feature of this solution is that, on switching to zig-zag stitching, from, say, the extreme left-hand or right-hand neutral setting of the needle-bar guide, a pattern is obtained of the kind solved shown in Figs. 2.65 and 2.66.

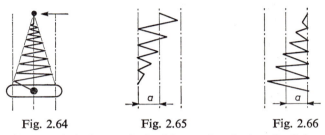

Fig. 2.64   Fig. 2.65   Fig. 2.66

Fig. 2.64. Different stitch shapes obtained by varying the length of zig-zag stitches.
Fig. 2.65. Stitch shape obtained by reducing the stitch size ($a = 0.5$ zig-zag length).
Fig. 2.66. Stitch shape obtained by enlarging the stitch ($a = 0.5$ zig-zag length).

The mechanism derives its motion from the eccentric cam (1) (Fig. 2.67). This arrangement has been adopted to avoid any sideways shift in the feed during the up-and-down movement of the needle. The needle-bar guide is driven at point A by member (2). The position of point B is determined amongst other things by that of point B_0. The motion of point A of member (2), projected in the vertical plane, is rectilinear, point A being a pin-joint on the needle-bar guide, which in this system can oscillate in a horizontal plane about a fixed centre of rotation (Fig. 2.62). By means of what is known about the motion of points A and B, we now determine the instantaneous centre P of the motion (Fig. 2.68a). If we turn control (3) in an anti-clockwise direction, the instantaneous centre moves towards A, until A and P coincide, whereupon the motion of A in the only direction permitted by eccentric (1), at once becomes zero, which means that the zig-zag motion also becomes zero instantaneously.

Throughout one full revolution of the eccentric (1), the motion is then such that the instantaneous centre (see Volume 1, Section 3.5) is restricted to a very short, closed curve, with A as a point on this. The result is that the residual zig-zag motion is negligible.

The neutral setting of the needle-bar guide, that is, the position it assumes when the zig-zag deflection is zero, can be adjusted by moving point B of member (2) horizontally, as can be done by means of the manual control (4) (Fig. 2.68b).

It will be seen that, during this adjustment, the vertical component of the displacement of B_0 is negligible, so the displacement of the neutral setting

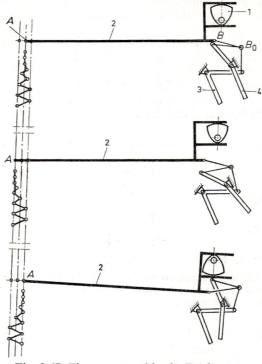

Fig. 2.67. Zig-zag control in the Dürkopp.

1 = eccentric. 2 = driving member for needle-bar guide. 3, 4 = manual controls

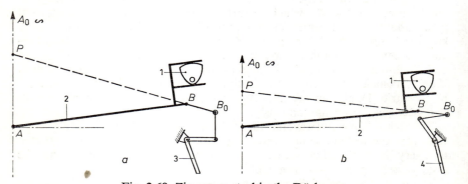

Fig. 2.68. Zig-zag control in the Dürkopp.

a. Adjustment of zig-zag deflection.   *b*. Adjustment of neutral position.

1 = eccentric.   3 = manual control for adjusting the zig-zag
2 = driving member for needle-   deflection.
    bar guide.   4 = manual control for adjusting the neutral
                                            setting.

by means of control (4) does not affect any zig-zag deflection already selected by control (3).

On the other hand, manipulation of control (3) does affect the neutral setting selected by control (4), since the path of point B_0 does include an appreciable horizontal component. The advantage of this is that it enables the neutral setting to be shifted in step with the adjustment of the zig-zag width, in such a way that larger zig-zag deflections, starting from an extreme left-hand or extreme right-hand neutral setting, produce a pattern which does not extend beyond these extremes in the left-hand direction, or right-hand direction, respectively (Figs. 2.65 and 2.66).

By means of this two-way continuous zig-zag control device, different decorative stitches can be produced.

To regularize them, a cam driven by the machine has been adopted as a means of obtaining the continuous adjustments needed for the zig-zag pattern. Some modern electric sewing machines are equipped with such a control device (Necchi, Elna, Pfaff). The Elna Supermatic is the first machine to incoporate a central control governing at one and the same time, both the variable zig-zag width and the variable feed, in both directions.

REFERENCES

[1] E. RENTERS, *Der Nähmaschinen-Fachmann*, Bielefeld, 1953.

[2] K. NICOMAY, *Getriebetechnische Aufgaben an Nähmaschinen*, Z.V.D.I. 96, 11/12 (15-4-54) p. 363.

[3] J. ZIEGLER, *Handbuch der Nähmaschine*, Aachen, 1953.

[4] H. APPELT, *Der Nähmaschinen und Spezial-Nähmaschinen in der DDR*, Leipzig, 1953. Der Nähmaschinen-Spezialist, Leipzig, 1954.

[5] *Bernina mit wesenlichen Konstruktions-Verbesserungen*, Deutsche Näh-maschinen Zeitung DNZ 84, (1963) 6, p. 37.

[6] H. GEITZ, *Die Schubkurbel als Nadelstangenantrieb bei Nähmaschinen*, Industrie Anzeiger 89 (1967) 34, p. 716.

[7] *Vorrichtung zum automatischen Heften*, DNZ-international/Die Nähmaschinen Zeitung 88 (1967) 2, p. 20.

[8] W. MEYER ZUR CAPELLE, F. J. GIERSE and B. SCHULER, *Massenkräfte in Nähmaschinengetrieben*, Feinwerktechnik 68 (1964) 6, p. 217.

[9] G. WUTHE, *Probleme beim Vernähen von Synthetikmaterial*, Deutsche Näh-maschinen Zeitung DNZ 86 (1965) 11, p. 20.

[10] TAVARO S.A. GENÈVE (Elna fabriek), *Funktionsweise des Elna-Para-Bloc-Greifers*, Deutsche Nähmaschinen Zeitung DNZ 86 (1965) 6, p. 46.

[11] *Technische Aenderungen an der Borletti-Nähmaschinen*, Deutsche Nähma-schinen-Zeitung DNZ 84 (1963) 12, p. 36.

[12] W. RENTERS, *Der Nähmaschinen-Fachmann*, 1957, Bielefeld, p. 89–106.

[13] W. RENTERS, *Die Nähmaschine in Schule und Haus*, Heidelberg, 1958, p. 43, 59, 57.

[14] A. G. A. KRUMME, *Mechanische Technologie voor de Confectie-industrie*, Misset, Doetinchem, 1966.

[15] C. E. FISCHER V. MOLLARD, *Untersuchung des Nähvorgangs bei Industrie-Schnell-Nähern*, Feinwerktechnik, February, 1967, p. 64.

[16] F. SCHMETZ, *Taschenbuch der Nähtechnik*, Herzogenrath, 1964.

Chapter 3

Optical Engineering Designs
JAC. v. NOORD

3.1 Introduction

This chapter deals with the mechanical structures and mechanisms employed in optical instruments. Apart from the design details, applicable to all instruments and apparatus, and the importance of using standard components, instruments employing lenses, prisms, mirrors, etc., have to meet special requirements. The optical calculations were discussed in Volume 1, Chapter 6.

The results of scientific experiments depend very much on accuracy of measurement, which in turn is governed very largely by two important factors: the measuring instruments employed, and the ability to reproduce the measurements.

Stability of the mount and proper functioning of the adjusting mechanisms should receive particular consideration in the design of optical apparatus.

3.2 Essential design requirements

3.2.1 Construction: fastening the optical system

The requirements for fastening optical components can be divided into two parts:

The components must be so arranged and fastened that the position and direction of the optical axis meet the desired standard of accuracy and the relative distances of the optical elements on this axis correspond with the calculated dimensions and tolerances.

The stability of the structure as a whole must not be disturbed whilst the instrument is in use.

Possible causes of disturbance are: deformation brought about by external forces; material stresses due to changes in temperature; play in moving parts; or vibration. The stringency of the stability requirement may be limited by such considerations as permissible weight and shape of construction.

This is best illustrated by examples, as follows:

A microscope has been carefully constructed to ensure that the object can be adjusted to the correct distance from the objective, at which a sharp and magnified image of the object is observed through the eyepiece.

So that this adjustment can be made correctly, the objective and the eyepiece are assembled in a microscope tube. The object is placed on a stage which can be moved in a plane at right-angles to the optical axis by means of a mechanical stage.

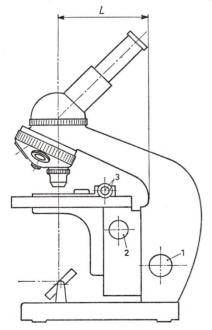

Fig. 3.1. Microscope.
1 = coarse-focusing knob.
2 = fine-focusing knob.
3 = knobs for mechanical stage.
L = free arm length.

Any flexure of the stage or the tube holder or support will cause a change in the selected distance. When this flexure is caused by a constant load, that remains unchanged as long as the microscope is in use, it will not affect the observation adversely.

But, if even a slight change of load, as a result of adjustment or manipulation, is enough to cause bending, the instrument cannot be used, particularly for relatively strong magnification. In this case, disturbing vibrations can also prove very troublesome.

The focusing adjustment of a microscope with 250 × magnification is about 0.5 μm. A change of this magnitude in the working distance results in an indistinct image [2].

When designing the instrument, it is necessary to ensure that the distance between the object and the objective remains unchanged after focusing, that is, throughout the actual observation. The tube holder must be attached to a solidly-constructed foot, or stand, whose arm L should not be longer than is strictly necessary for manipulating the object.

The control knobs for focusing and for moving the object must be arranged so as to avoid, or at any rate minimize, flexure of the stand due to the forces exerted on them (Fig. 3.1).

Photography without a stand is only possible with short exposure times and if the camera is held motionless during the exposure.

For longer exposure times, a stand or tripod and a cable-release are essential. Great stability is equally essential for all optical instruments. Although many of the optical instruments discussed here weigh very little, (not more than a few kilogrammes) and weight has practically no effect on the mechanical strength of the stand, base plate, stage, etc., their usefulness depends very much on a good, reliable and stable support. The greater the magnification obtainable with the instrument, the more important is stability.

3.2.2 Adjustment and alignment

Calculations of lenses, mirrors, prisms, diaphragms and the like in optical instruments are made with the aid of geometrical optics.

In designing such instruments, the components to which these calculations relate have to be given dimensions and tolerances on the drawing. Nevertheless, the design, especially of very accurate instruments with compound optical systems, must leave scope for correction during assembly. This alignment has to be carried out once and for all during final assembly, after which, and only then, the instrument is ready for use. Special apparatus and tools are often required for the purpose.

In prism binoculars, the optical axes of the two telescope-members must run parallel. This adjustment is usually made at the objective-end. If the optical axes of the individual telescopes are not parallel, double images are formed. Vertical and horizontal differences between them are called vertical and horizontal deviations, respectively (see Table 3.1). Two fixed collimators are used for this adjustment (Fig. 3.2). The two collimators with a datum mark, and two telescopes with a measuring

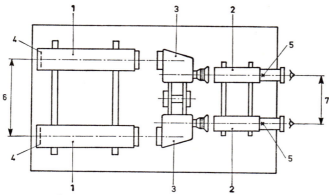

Fig. 3.2. Equipment for adjusting binoculars.

1 = collimator.
2 = telescope.
3 = binoculars.
4 = datum mark.

5 = scale.
6 = distance between objectives.
7 = distance between eyepieces.

eyepiece, are mounted immovably on a frame. Their optical axes are all aligned exactly parallel. The binoculars to be adjusted are placed in a holder between the collimators and the telescopes. The distance between the telescopes equals the eye distance of the binoculars, e.g. 65 mm. The distance between the collimators is equal to the distance between the binocular objectives, which corresponds to the distance between the eyes.

If the binocular axes are not parallel, the image of the datum or index marks will be displaced. This displacement can then be read from the scale in the telescope [3].

TABLE 3.1

Permissible angular deviation in binoculars

Binocular magnification	Horizontal		Vertical
	Convergent	Divergent	
3 ×	70′	35′	17′
6 ×	28′	14′	7′
8 ×	20′	10′	5′
10 ×	15′	8′	4′
12 ×	12′	6′	3′

An entirely different correction, also made during assembly, is the adjustment of the revolving nosepiece of a microscope. Here, the screw-on faces of the objectives must in turn be normal to the optical axis, whilst the distances to the tube of the microscope must be the same. To accomplish this, these faces are finished to the same size after assembly, a fact that must be taken into account in the design (Fig. 3.3).

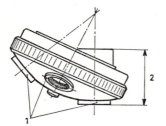

Fig. 3.3. Microscope nosepiece.

1 = faces to be finished.
2 = distance between parallel faces.

The screw-on faces of the microscope objectives are likewise finish-turned after assembly, whereby the position of the optical axis is corrected if necessary. At the same time, the optical length (the distance from object to screw-on face) is adjusted to conform to the standard dimension. These identical lengths are necessary, for objectives of different power to be used by the nosepiece, without changing the coarse adjustment. Only the fine adjustment should be used.

Of course, these and other similar adjustments constitute the final phase of manufacture.

The instrument should not be altered in any way when in use, since the exact data, and the auxiliary equipment and special tools required for the purpose, are rarely available.

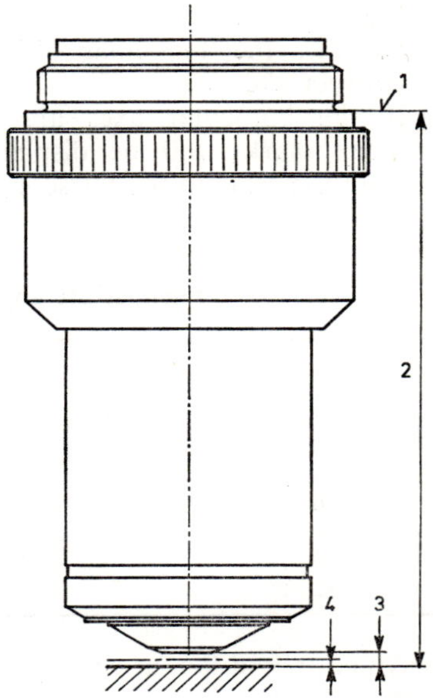

Fig. 3.4. Microscope objective.

1 = screw-on face.
2 = distance from tube to object
    (standard dimension).
3 = free object distance.
4 = cover glass thickness.

3.2.3 Focusing

Optical instruments are used so that an object can be imaged or observed. In the process, certain adjustments have to be made so that the instrument can fulfil this purpose. The image formed may be larger, or smaller, than the object itself.

The adjustments fall into two groups: those following, and those in a plane normal to, the direction of the optical axis. Any displacement co-directional with the optical axis changes the distances established in the optical plan.

In the case of a simple lens, with a given object distance and a given image distance, changing the one distance produces a change in the other, as also in the magnification. Likewise, the same results can be obtained by changing the distances between the optical elements of a compound optical system (Figs. 3.5 and 3.6). Thus, the image of an object can be focused and magnified to suit the particular purpose.

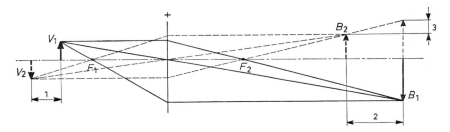

Fig. 3.5. Path of rays through a thin lens.
1 = change in object distance.   3 = difference in size of the image.
2 = change in image distance.

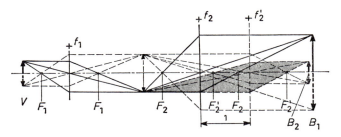

Fig. 3.6. Path of rays through two thin lenses.   1 = displacement of one lens.

Through an adjustment in the plane normal to the optical axis, a different point of the object can be brought on to this axis. In this way, different parts of the object under examination can be brought into the image, or the optical axis can be brought into exact alignment with the centre of the object.

(a) Adjustment along the optical axis

This brings the image into sharp focus and regulates its size. The length of adjustment required, and its accuracy, depend on the type of instrument, whilst the depth of focus is an important factor.

(b) *Classification*

Optical instruments may be divided into two categories, depending on the manner in which the image is perceived:

- Instruments that form an image on the retina of the eye, in which case the eye should be regarded as an essential part of the instrument, with its lens and iris included in the direct path of rays. The accommodation of the eye and its pupil width also have a part in this. Instruments in this category include the magnifying glass, spectacles, the microscope and various binoculars and telescopes.
- Instruments that form an image in a focal plane or on a screen. They include the camera and the projector.

A further classification can be made, according to the position of the object relative to the instrument.

- The object can occupy a fixed position on or in the instrument, as in the case of the microscope, the projector and the copying camera.
- The object can be completely detached from the instrument at distances varying from near-by to infinity. This group includes the magnifying glass, spectacles, cameras in the ordinary sense, binoculars and telescopes.

The diversity of the instruments gives scope, of course, for a number of design solutions with regard to the adjusting mechanisms.

(c) *Optical axis*

A concept of importance to all optical instruments is the *optical axis*. It is this which, amongst other things, dictates the form the instrument should take, from the points of view of mechanical construction and design.

There are two alternatives:

- The optical axis is within the mechanical guide. This produces a compact and often stable structure, but limits the dimensions. The necessary adjustments can be made by means of:

    A sliding tube, which can be adjusted by hand, with or without a clamp.
    A sliding tube, adjusted by means of a rack or leadscrew, as in a focusing telescope, astronomical telescope or micro-optical bench.

- The optical axis is outside the mechanical guide. The structure is then much less compact and more attention must be given to stability. In general the dimensions are subject to fewer limitations. Adjustments can be made as follows:

    A carriage, beam or longitudinal guide with slides, which can be moved by hand and clamped in the desired position.
    A carriage shifted by means of a leadscrew, a rack or a special reduction gear, for accurate adjustments. For example: microscope, copying camera, optical bench.

Generally, all these devices are designed to bring an image into sharp focus. The guides must move without juddering or jolting, and the clamps must grip reliably. Backlash must be eliminated where necessary.

These matters will be discussed more fully in the following paragraphs.

3.2.4 Dismantling and overhaul

It will now be evident that due care must be taken in dismantling optical instruments. The likely difficulties in reassembling and adjusting the instrument should be noted carefully beforehand. Even in the planning stage, the designer must bear in mind that overhaul may be needed, to eliminate, for one thing, any play caused by wear. Dismantling may also be necessary to clean the instrument or to replace damaged parts.

How these devices are constructed depends on the accuracy required, important factors being: the climate, workroom and circumstances in which the instrument is to be used. From this point of view, there is, of course, a lot of difference between, say, a telescope for field conditions outdoors and one permanently installed in an air-conditioned chamber.

It is also important to consider who will have to dismantle and overhaul the instrument. This work may be left to the maker, because he, above all, has the necessary knowledge and equipment.

Again, it may be that repairs are carried out by an experienced craftsman with a properly-equipped instrument workshop at his disposal. And there are users quite capable of carrying out the necessary overhaul themselves, without assistance.

3.3 Components

The material properties of the glass of lenses, mirrors, prisms, windows, filters and so on, must be taken into account in deciding how to fasten optical components.

The difference in material constants of glass and the metals used to make the mounts are generally sufficiently large to merit due consideration during design (Table 3.2).

TABLE 3.2

Material	Coefficient of linear expansion at 20°C ($\times 10^{-6}$/°C)
Mirror glass	9
Quartz glass	0·6
Opt. glass	5–10·5
Steel	10–12
Brass	19
Al. alloy	23
Fernico	6
Invar	1
Molybdenum	5

Incorrect fastening of the optical system can cause the glass member to be broken or damaged. Also, it may happen that the optical properties are affected by the internal stress in the glass or by a change in the shape of the surfaces (see Volume 2, Chapter 3).

The dimensions and data needed to construct optical instruments are given in the optical design [4].

The following sizes and data should be specified in this "optical scheme":
Dimensions of the glass members (lenses, prisms, etc.).
The effective apertures required.
The relative distances.
The position and dimensions of diaphragms.
The position in the instrument as a whole (object and image distances).
Tolerances on the aforementioned dimensions.
Adjustment and correction facilities required.
Physical data of the glass.
Particulars of polishing and inspection of the optical surfaces (see DIN 3140).
Particulars of the finishing of the optical surfaces (coating, metallizing).

3.3.1 Lenses

The object of fastening lenses in a metal mount is to anchor them in a fixed position in the instrument. This may be done with a simple lens, or with a compound, cemented system. The correct location of the lens depends on the direction and position of the optical axis. After the optical surfaces have been polished, the lens is ground to the exact diameter required, by the optical grinder. In this process, also called centring, the axis of the cylinder is made to coincide with the optical axis of the lens [4,8]. Bevelling can be done in the same operation (see Volume 4, Chapter 2). The bevelled edges, or facets, are important both in the manufacture of the lenses and in the subsequent fitting in the mount. By bevelling the lens, the chance of damaging the edges is greatly reduced. When the lens is fastened in the mount by spinning it in, the facet has a structural function.

Figure 3.7 shows an example in the form of a biconvex lens. With the principal dimensions given here, together with the data from the optical diagram, it is possible to design a suitable mount.

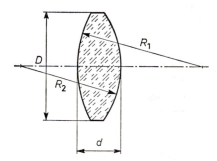

Fig. 3.7. Dimensions of biconvex lens.

(a) *The lens seat*

There are several ways of retaining a lens in a metal mount or setting. Practically all of them involve pressing the lens against a solid rim, or lens seat. This seat can be formed in three different ways, as shown in Fig. 3.8.

The first (Fig. 3.8*a*), whereby the seat mates exactly with the spherical shape of the lens, is purely theoretical and impracticable in a metal mount.

The second method (Fig. 3.8*b*) is to press the lens against a tangential face, whereby the seat is tapered outwards by turning. In so doing, the tangent circle must be kept far enough from the edge of the lens to avoid setting up dangerous stresses here, where the lens is vulnerable.

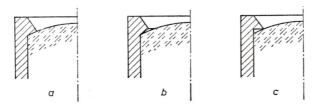

Fig. 3.8. Lens seat.   *a*. Convex shape.   *b*. Tangent plane.   *c*. Straight.

The risk is that the edge will splinter, which is almost certain to happen if it is not given a protective facet. The third method is to give the seat a straight bore (Fig. 3.8*c*) which, like the previous method, also involves a circle at which the mount touches the spherical shape of the lens. However, this circle is as far as possible from the edge of the lens and is therefore the least likely to damage it.

The place where the lens touches the seat is determined exactly, since the effective aperture of the mount is also the circle tangent to the spherical surface of the lens. The calculation of the lens position in the mount is a simple one (Fig. 3.9).

A further advantage is that the lens face can be cleaned more thoroughly and easily, because there is no space between lens and mount into which dust and hairs may find their way. This is particularly important for the

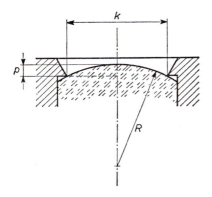

Fig. 3.9. Determining the position of the lens in the mount.
k = effective aperture.
p = height of segment.
R = radius.
Formula: $\left(\dfrac{k}{2}\right)^2 = (2R - p)p.$

lens faces on the outside of an instrument, which are directly exposed to possible contaminations from outside, and therefore have to be cleaned from time to time.

The rim must be accurately made and free from burr. The radial clearance between lens and mount depends on the type of instrument and the conditions under which it is to be used.

The differences in coefficient of expansion mentioned earlier should be taken into account. One or two examples of usual clearances as a guide are[2]:

Microscope objectives ~ 0.01 mm (ISO F7/h6).
Telescope objectives 0·05–0·1 mm (ISO F8/h7).
Condenser lenses in illuminating systems up to 2 mm.

(b) *Fastening the lens*

Whether the lens should be retained in the mount by a permanent joint or a detachable fastening depends on its position in the apparatus as a whole and the possibility of future overhaul. Where the joint is permanent, it is nearly always desirable for the lens faces to remain accessible when the instrument is dismantled. Therefore, no two separate lenses should be mounted permanently in the same barrel, for it would then be impossible to clean the two glass surfaces facing each other. This would inevitably become necessary in the long run, particularly for instruments used in a humid or dusty environment.

Permanent fastenings are not used for lenses of more than 30 mm diameter.

The fact is that permanent mounting is a particularly attractive proposition for the smaller lenses, in view of the limited space available, and is often the only possible method in complex optical systems.

Lenses of less than 6 mm diameter are sometimes cemented into the barrel, by making a cement joint between the cylindrical surfaces of lens and barrel. During this process, the lens must be pressed against the lens seat (see Volume 2, Chapter 8 and Volume 7, Chapter 4).

Occasionally, for very accurate optical systems, precise centring is ensured by re-turning the mount with the lens cemented into it, so that the

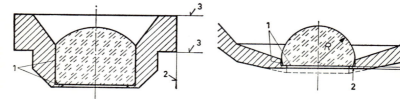

Fig. 3.10. Mount with lens cemented in.
1 = cement face.   3 = mating face.
2 = centring face.

Fig. 3.11. Front lens mount of microscope objective.
1 = cement face.
2 = material to be turned off.
d = thickness of mount rim.

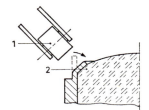

Fig. 3.12. Spinning rim.
1 = spinning tool.
2 = lens facet.

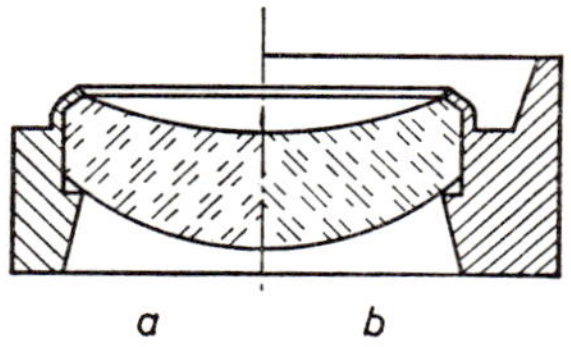

Fig. 3.13. Lens forced in.
a. Flat mount.
b. Mount with upright edge.

optical axis coincides with the axis of the finish-turned cylinder, and any mating flange is exactly at right-angles to this axis (Fig. 3.10).

Re-turning the mount, after the lens has been cemented in, is also necessary when the mount material is so thin (i.e. 0·02 to 0·03 mm) that it cannot be turned accurately without the lens fixed in it. An example of this is the front lens of a powerful microscope objective (Fig. 3.11). Because the effective object distance is so short here, the rim of the mount cannot be made any thicker. Obviously, the mount cannot be machined again after this turning process, so it is made of a tough metal alloy. With larger lenses, the differences in the materials would give rise to undue stresses in the cemented face.

Another method often used is spinning the lens into the mount. This involves grinding a facet at the edge of the lens and then spinning-over the rim of the mount on to it .[6] A spinning tool or roller is used for the purpose (Fig. 3.12). During this process, the lens must be pressed firmly against its seat, and on a special spinning lathe this can be done by tilting the spindle slightly. The spinning rim on the mount must be thinned down to a suitable height on the lathe; the rim thickness depends on the diameter and material of the mount (0·1 to 0·3 mm). This thin rim should extend far enough down to bring its junction with the thicker part of the mount below the start of the lens facet (Figs. 3.13*a* and *b*).

The depth of the recess in the mount calls for careful consideration, particularly in the case of small precision parts. The spinning rim for a 0·2 mm

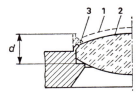

Fig. 3.14. Depth of recess in mount.
1 = lens of maximum thickness.
2 = lens of average thickness.
3 = surplus stock to be turned off.
d = depth of recess.

Fig. 3.15. Lens with pressure washer.
1 = grooves.

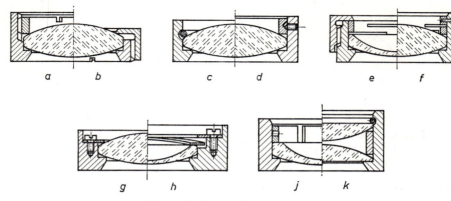

Fig. 3.16. Detachable lens mounts.

a. Threaded ring.
b. Nut.
c. Spring wire clip.
d. Pressure washer with three pointed
    screw.
e. Spring packing ring.

f. Spring retaining ring.
g. Pressure washer.
h. Pressure washer with corrugated
    (or wavy) washer.
j. Clamping bush.
k. Distance-piece between two lenses.

facet at 45° is less than 0·3 mm. The tolerances on lens thickness and recess depth combined may be of the same order of magnitude (Fig. 3.14).

In such cases, the mount must therefore be made deep enough for the "thickest" lens to be spun into it without difficulty. The result is that, when a thinner lens is inserted, there is too much material on the rim. This surplus must be turned off during the spinning operation. Lenses can also be spun-in with a metal ring between them and the rim (Fig. 3.15).

A seat similar to that already described is used where the lens fastening is not permanent. The lens is then retained in the recess by a detachable threaded ring, clamping ring, spring clip, etc.

Because of the stress in the glass, care should be taken not to exert undue force on the lens during assembly. When a completely stress-free optical system is required, for example, where polarized light is to be used, special provision must be made for this in the design. A few examples of detachable lens mounts are shown in Figs. 3.16*a* to *k* inclusive.

3.3.2 Prisms

The diversity of shapes to be found in this group of optical components has led to a range of altogether different fastening methods and structures.

The method chosen depends very much on the conditions under which the component is to be used. Entirely different requirements are imposed on the fastening of the prisms in prism binoculars, than on the fastening of the prisms in a liquid refractometer. The prisms in the binoculars are contained in a completely dust-proof casing: those in the refractometer are wetted by the liquid on test during each experiment and have to be cleaned thoroughly afterwards.

One or two general points to be considered in the design of prism mounts are[6]:

Clamping or cementing should be done without any risk of the glass member being deformed.

Stress in the glass must be avoided, and the difference in coefficient of expansion of the materials employed should not be large enough to cause such stress.

The prism must be locked in exactly the right position, so as to resist shock and jolting in use or when being conveyed from one place to another.

The bearing face must be adequate, without putting any strain on the fragile corners of the prism. The optically-active faces must not be used as bearing faces, unless provided with a metallized reflecting film.

Uniform distribution of the clamping force is obtained by interposing a soft material, such as cork, paper, aluminium, plastic, etc.

A simple prism mount is illustrated in Fig. 3.17. The prism is clamped in a resilient plate with a pivot riveted on. The bowed plate is well clear of the reflecting face, whilst its beaded edges clamp round the corner facets of the prism. Figure 3.18 shows the prism mount used in prism binoculars. The seat in the light-metal housing is made with an end mill. The apices of the 45° angles of the prism have been cut off semi-circularly, since the light rays

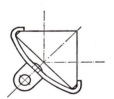

Fig. 3.17. Prism clamp.

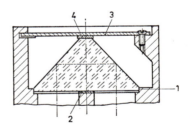

Fig. 3.18. Prism mount in binocular
housing.
1 = seat.
2 = shoe.
3 = clamping spring.
4 = cork pad.

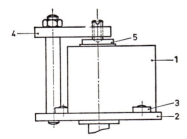

Fig. 3.19. Prism on rotary platform.
1 = prism.
2 = rotary platform.
3 = packing plates.
4 = arm.
5 = pressure plate with cork pad.

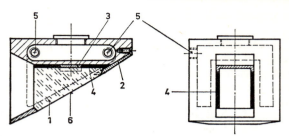

Fig. 3.20. Refractometer prism mount.

1 = prism.	4 = cement.
2 = locking plate.	5 = circulating duct.
3 = pressure plate.	6 = measuring surface.

do not pass through them and so they take no part in the image-forming process. The space thus saved is an important factor in the compact structure of the binoculars. The prism is clamped at the rib of the 90° angle, on which a flat face is ground. The prism rests on a thin plate of soft aluminium, mounted in the middle of the seat and retained against sliding. The clamping spring is faced with cork where it presses against the prism. The spring is attached to the housing by sliding one end into a groove and screwing the other end on to a lug.

Prisms in laboratory test apparatus, like the spectroscope, monochromator, etc., can be fastened in the manner of Fig. 3.19. The prism rests on a flat, rotary platform and is held in the correct position by one or two plates. A screw inserted in an arm attached to the platform thrusts a pressure plate down on to the top of the prism, with a cork pad between them.

Each of the pair of prisms in the Abbe refractometer is cemented into a mount. The cement must be impervious to the liquids tested and, for these measurements of the refractive index, the temperature of the liquid must be kept constant at a given value. This is accomplished by circulating water at a constant temperature through ducts in the mount. Figure 3.20 illustrates one of these prism blocks. To fill the gap between the measuring and illuminating prisms completely with liquid, the metal surfaces are slanted off slightly. The prism is placed in the mount and retained by a locking plate inserted afterwards. A pressure plate fitted behind the prism holds this in position until the cement is applied. The space between mount and prism is then filled with cement.

3.3.3 Mirrors

These components have only one optically-functioning surface. After being reflected, the light emerges from the same side that received the incident ray. The incident and reflected rays are at identical angles to the normal (perpendicular to the mirror surface) (see Volume 1, Chapter 6). Tilting the mirror through an angle α involves an equivalent angular displacement of the normal. The reflected ray is thus rotated through 2α relative to the incident ray (Fig. 3.21). The space behind the mirror is of no importance optically and this gives scope for other methods of fastening than the aforementioned lens mounts. The mirror-mount in a luminous-spot galvanometer is a simple example of a fastening at the back of a mirror (Fig. 3.22). Here, the mirror is cemented direct to the coil or strip, a method of fastening which in this case is essential since, to save weight, there is no extra mount and the mirror is made as light as possible.

Another example is the fastening of a concave mirror and a flat mirror in one mount, as in microscope illuminators for translucent objects. The mirrors are spun-in like the lenses discussed in Section 3.3.1. The drawback, that two glass surfaces facing each other cannot be cleaned, does not apply here, and this construction affords complete protection to the two vulnerable mirror-surfaces (Fig. 3.23).

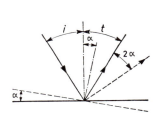

Fig. 3.21. Path of rays at a flat mirror.
  i = angle of incident ray.
  t = angle of reflected ray.
  α = angle of rotation of mirror.

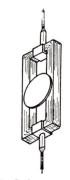

Fig. 3.22. Galvanometer mirror.

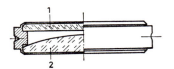

Fig. 3.23. Microscope mirror.
  1 = flat mirror.
  2 = concave mirror.

The requirements for mounting mirrors in precision measuring apparatus and instruments is altogether different. Deformation of the reflecting surfaces must be avoided here, adjustments must be carried out with the utmost precision and the fastenings must be very stable. The reflecting surface will be deformed if the mirror clamped in position is subjected to a flexural load. The mirrors in such precision instruments have the reflecting surface at the front, so the thickness of the glass is unimportant optically, since it is not part of the optical path. On the other hand, the thickness of the glass is a very important factor in the manufacture of accurate optical surfaces, because flexure during processing and mounting must be avoided. As a general rule, the thickness should be one-tenth of the length.

Mirrors for optical instruments are made mainly of glass, with a metallized reflecting surface. This reflecting film can be obtained by vapour deposition in vacuum or by chemical precipitation (see Volume 6, Chapter 2). These glass mirrors are used mainly in instruments where exact images are required, for measurements in the visible and ultraviolet spectra.

Metal mirrors are manufactured for reflectors in lighting equipment.

The mirrors are pressed by a process that does not involve chip-forming, and are then polished mechanically or electrolytically. Such all-metal mirrors can be given an aspherical shape and are relatively less vulnerable.

Reflecting surfaces of a very high standard can be made by machining metal with a diamond-tipped tool on a stable lathe specially equipped for the purpose (see Volume 5, Chapter 1).

The mirror fastening must be so designed as to rule out any risk of buckling under load. The positioning of the supporting and fastening points must meet the following requirements:

The mirror must assume a fixed position and be incapable of shifting.

The three supporting points must be exactly opposite the three clamping points, so that the mirror is only subjected to compressive forces (Fig. 3.24).

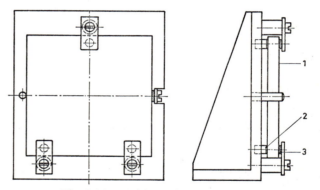

Fig. 3.24. Precision mirror mounting.
1 = reflecting surface.   2 = bearing point.   3 = pressure point.

3.3.4 Glass plates

Clamped, flat glass plates are used in optical instruments as a base for the object to be observed, or as a screen on which to produce the image. The position of the glass plate is specified in the optical plan as the object plane or the image plane.

An example of the glass plate as a base for an object is the graduated glass plate in an eyepiece micrometer. The frosted glass in a projection microscope, or the frosted glass and the photographic plate in a plate camera, are examples of the plate on which an image is produced.

They are in a plane perpendicular to the optical axis. In one or two other instruments, the plate has to be slanted. For example, the photographic plate in the spectroscope camera, and in the technical camera with perspective correction.

Either permanent fastening or quick changing of the glass plate may be called for. In both cases, the fastening should hold the plate immovably in the correct position without bending it. To achieve this, the pressure points of the springs must be made to coincide exactly with the supporting points of the plate. In the case of instruments permanently installed in enclosures equipped with a vacuum suction line, a suitable fastening can be obtained by using vacuum suction to draw the glass plate tightly into position.

Figure 3.25 illustrates a vacuum suction plate which can be used to fasten a photographic plate or film to a flat, porous surface (see Volume 3, Chapter 5).

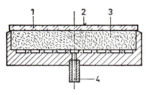 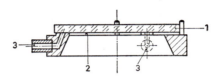

Fig. 3.25. Suction plate for glass or film.
I = glass plate or film.
2 = photosensitive coating.
3 = porous material (sintered bronze).
4 = hose coupling for vacuum line.

Fig. 3.26. Suction frame for glass plate.
1 = glass plate.
2 = photosensitive coating (image plane).
3 = vacuum duct.

In this method, the photosensitive coating faces away from the suction surface. The thickness of the base (glass plate or film) is a possible source of error in exact focusing.

The holder shown in Fig. 3.26 holds the glass plate by suction at three points, which also serve as supports. This permits mating with the photosensitive side of the plate, so that the thickness of the glass does not affect the focusing of the image.

3.3.5 Glass filters

Light filters placed in the path of rays of an optical system absorb some of the rays.

There are: colour filters, ultraviolet filters, heat-absorbing filters (or heat shields), polarizing filters (or screens) and so on.

The position of the filters in an optical system depends on the purpose; the positioning need not be very accurate, but the polarizing filter must be arranged so that it can be faced in the correct direction.

The infrared rays absorbed by the heat-absorbing filter raise the glass to high temperatures, so the filters must be mounted with a substantial clearance.

Where a ventilator is used to provide forced cooling, the heat-absorbing filter must be given the capacity to lose the heat uniformly. Fig. 3.27 shows the positioning of the filter mounted in the radiation screen of a projection lamp.

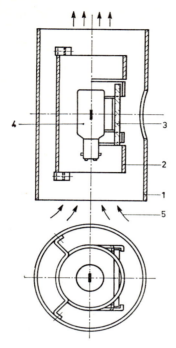

Fig. 3.27. Lamp housing with filter.
1 = housing.
2 = radiation screen.
3 = heat-absorbing filter.
4 = lamp.
5 = air current.

3.3.6 Diaphragms (or stops)

The diaphragm (or stop) is an optical component which is part of the optical path but is not made of glass. It has two functions: to limit the light beam, and to keep stray light rays out of the beam. The diaphragm has no part in the forming of the image by refraction or reflection of the rays, but this does not make it any less necessary and important to the proper functioning of the optical system, since the size of the aperture and the position in the system are critical factors governing the quality of the image.

These dimensions are specified in the optical plan. Figure 3.28 illustrates the functions of the stops in a simple camera.

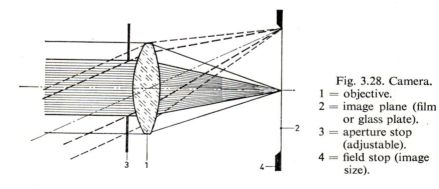

Fig. 3.28. Camera.
1 = objective.
2 = image plane (film or glass plate).
3 = aperture stop (adjustable).
4 = field stop (image size).

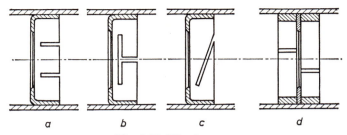

Fig. 3.29. Fixed stops.

a, b, c. Ring, pressed from sheet material, with spring· clip
d. Flat ring clamped-in.

The stops are mounted in a plane normal to the optical axis, and Fig. 3.29 shows different versions that can be used when a fixed stop is required. The ring, formed of sheet material, is clamped in the tube by a cylindrical spring clip. The rim of the aperture should be as thin as possible and can be turned off at an angle if necessary. The sharp rim is then usually positioned to face in the direction in which the light is travelling, that is, away from the light source. However, it faces the opposite way in stops which limit the field of view (field stop in an eyepiece). In a measuring eyepiece, the graticule lies against the field stop. The graduated side adjoins the sharp rim of the aperture, so that both are seen in sharp focus (Fig. 3.30).

In apparatus where the apertures have to be adapted to different optical conditions, either exchangeable fixed stops or continuously variable stops are employed.

The exchangeable stop can be made in the form of a simple ring or plate with the desired aperture; it is then placed in a holder and centred. An extension of this principle is a disc or strip with several different apertures which are brought into the optical path in turn.[1]

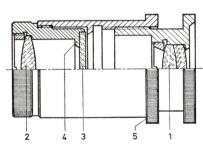

Fig. 3.30. Measuring eyepiece.
1 = eye lens (adjustable).
2 = field lens.
3 = graticule.
4 = field stop.
5 = rim of tube.

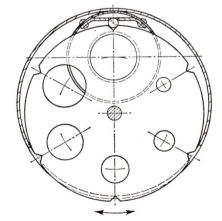

Fig. 3.31. Rotating stop.

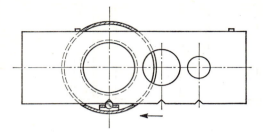

Fig. 3.32. Sliding stop.

For the rotating stop (Fig. 3.31), the apertures are made in a disc rotating on an axis parallel to the optical axis. A snap device ensures correct positioning of the aperture selected. A segment can be used instead of the disc, where only a few apertures are required.

The apertures in a sliding stop are selected by moving a strip in a straight line, whilst correct register of the aperture is ensured by the guide and the snap device.

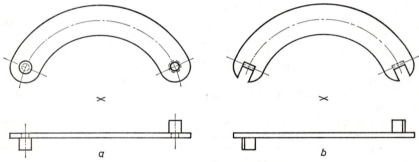

Fig. 3.33. Leaf of iris diaphragm.
a. With studs.   *b*. With erected tabs.

A continuously variable aperture can be obtained with an iris diaphragm. The aperture is defined by sickle-shaped blades, or leaves, with studs on opposite sides at each end (Fig. 3.33*a*) or with two erected tabs (Fig. 3.33*b*). The studs or tabs at one end are pivoted in a fixed ring, the diaphragm housing, whilst those at the other end slide in slots cut radially in a rotary ring (Fig. 3.34). The number of leaves governs the shape of the aperture: this is round at its widest and polygonal at its narrowest. The smallest aperture obtainable depends on the thickness and the number of the leaves. To avoid damage to the edges and the studs, a stop must be provided on the rotary ring at a point before this smallest aperture is reached.

A square aperture can be obtained by means of two strips moving in opposite directions (Fig. 3.35). This invariably-square aperture can be varied continuously from its maximum to completely closed. Because this involves only slight friction, little force is required to move the system, so it is popular for automatic stops.

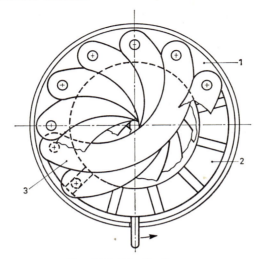

Fig. 3.34. Iris diaphragm.
1 = housing.   2 = ring.   3 = leaf.

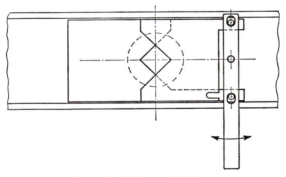

Fig. 3.35. Cat's-eye diaphragm.

3.3.7 Slits

Spectroscopic instruments require apertures in the form of slits, namely, as an entrance slit for light in the spectroscope, and as an exit slit as well in the monochromator.[7] The results of measurements with these instruments depend very much on the precise adjustment and finish of the slits. The effective resolving power (purity of the spectrum observed) is governed in part by the width of the slit, ranging from a few μm to about 1 mm. The following types of slit can be identified, depending on the uses and accuracy required:

(a) Firstly, according to the *adjustability*:
Fixed slit.
Unilaterally adjustable.
Bilaterally adjustable.

(b) Secondly, according to the *shape of the aperture*:
  Straight aperture.
  Curved aperture.

The classification in terms of adjustability can also be augmented as follows:

(1) By an adjustment of the slit height.

(2) The coupled or tandem slit.

The following points are worth noting with regard to the design and manufacture of slits.

The finished sides of the slit must be sharp and clean-cut.

The parallelism of the aperture must satisfy stringent requirements, subject to a tolerance of 1 or 2 µm.

When closed, a slit must not admit any light at all: for this reason, any damage to the sharp edge and any dirt trapped in the slit will be visible at once.

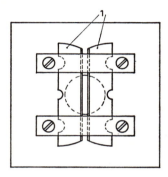

Fig. 3.36. Simple fixed slit (temporary).
1 = razor blades.

It will be evident from this that the blades or leaves forming the slit must be made of a special material, particularly for apparatus in which intense illumination is required, where a powerful beam of light is concentrated on the entrance aperture. Stainless, non-magnetic chromium nickel steel is very suitable for the purpose.

The fixed slit is adjusted to a predetermined width on assembly and is not altered during measurements with the instrument. Exchangeable slits of different widths are sometimes used in instruments of this kind.

In a temporary test set-up, two razor blades can be clamped to form a straight slit (Fig. 3.36) (1).

One member of the unilaterally adjustable slit can be moved relative to the other, which is mounted in a fixed position. A drawback of this method is that the slit centre is also displaced by the equivalent of half the slit adjustment. Not so with the bilaterally adjustable slit, for the two members move symmetrically about the centre of the aperture.

Very stringent requirements are imposed on the guide controlling the motion of the two slit members. The adjusting mechanism must not press the sharp edges together too forcibly, since this might damage them. To govern this force, the two members are brought together by a spring, whose pressure is specific and unvarying.

The adjusting mechanism then has to open the slit against the pressure of the spring.

The travel of the slit members can be made purely diametral by means of a straight parallel guide. This construction can also be used for a curved slit aperture. The centres of curvature are on a line parallel to the direction of travel (Fig. 3.37).

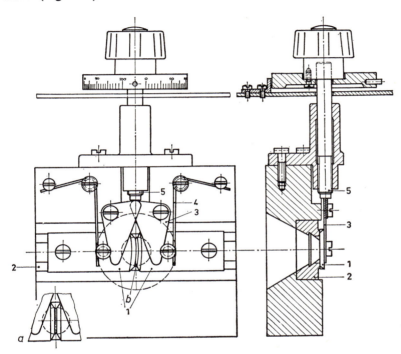

Fig. 3.37. Adjustable slit (straight parallel guide).

1 = members forming the slit.
    a. Straight slit. *b*. Curved slit.
2 = guide.

3 = pressure piece (lever).
4 = spring.
5 = adjusting screw.

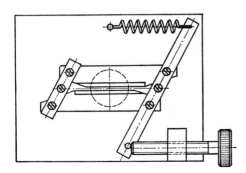

Fig. 3.38. Adjustable slit (rod guide).

Using a system of rods to shift the slit members (Fig. 3.38) confers motion not only across, but also in line with the slit. This mechanism can only be used for a straight slit aperture.

3.4 Components of the shift mechanism

Apart from the optical elements just discussed, all of which are placed in the optical path, other components are required to construct complete optical instruments. However, these are mechanical design elements used in the construction of precision instruments generally.

3.4.1 Guides (rectilinear)

Accurate guides are required to shift and adjust the optical objects, etc., and the shift obtained with them must be smooth, reproducible and without play. The structural design of such guides is governed by the displacement required and the load imposed on the members. A straight guide spanning only a few mm can be made with two flat springs mounted parallel to one another (Fig. 3.39). The surface A is shifted by the force F exactly parallel to surface B. Given a suitable choice for the ratio of spring length to the minor longitudinal displacement, the associated change of distance at right-angles to the above will not present any practical drawback.

The travel obtained with this construction is frictionless, rectilinear and free from play.

Ball and roller guides are suitable for travels ranging in length from some cm to some dm. Precision vee-grooves are machined in the hardened steel bars (Fig. 3.40).

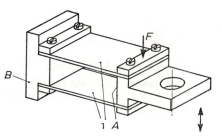

Fig. 3.39. Straight guide formed by parallel flat springs.
    1 = two identical flat springs.
    A = moving block, or holder.
    B = fixed block.

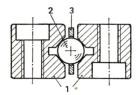

Fig. 3.40. Straight ball-guide.
    1 = vee-grooved guide.
    2 = ball.
    3 = cage.

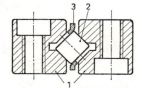

Fig. 3.41. Straight roller guide.
    1 = vee-grooved guide.
    2 = roller.
    3 = cage.

In these vee-grooves are placed several balls, held apart by a perforated strip corresponding to the cage in a ball bearing. The rolling friction of this guide is slight: if properly assembled, it ensures accurate travel, with scope for precision adjustment to within a few μm. Although the load capacity is ample for most instruments, it can be increased where necessary by using rollers instead of balls. The rollers are slanted in the vee-grooves (Fig. 3.41).

A few firms supply complete sets of guide bars in various sizes. These members are supplied complete with fixing holes and can be used, together with other accurate but easily-made parts, to construct a slide (see manufacturers' catalogues).

Instead of the hardened steel, vee-grooved bars, the following can be used to make a ball guide:

(a) Four hardened steel strips, clamped in grooves cut in two bars of unhardened material (steel, cast iron, brass, aluminium alloy, etc.) (Fig. 3.42). After the steel strips have been inserted, their edges are accurately dressed, to serve as a track for the balls.

(b) Two rectangular-grooved bars with steel wires pressed into the corners of the grooves by the balls. These four wires then constitute the ball track (Fig. 3.43).

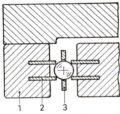

Fig. 3.42. Ball-guide with track formed by hardened steel strips.
   1 = metal bar with saw cut.
   2 = hardened steel strip.
   3 = ball cage.

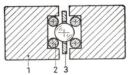

Fig. 3.43. Ball-guide with track formed by steel wires.
   1 = grooved metal bar.
   2 = steel wire.
   3 = ball cage.

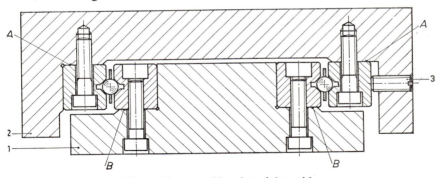

Fig. 3.44. Assembly of straight guide.
   1 = fixed guide bar.   A = mating face of top piece.
   2 = adjustable guide bar.   B = mating face of bottom piece.
   3 = grub screw.

Fig. 3.45. Dimensions of
the straight guide.
S = stroke or travel.
L = beam length.
K = length of ball track.
B = width of guide.

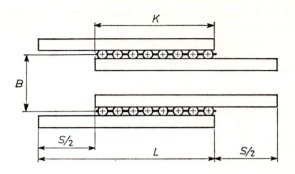

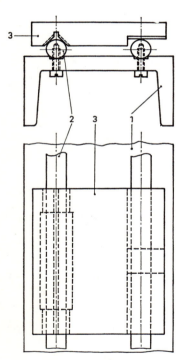

Fig. 3.46. Rod guide.
1 = channel.
2 = round steel bar.
3 = slide or carriage.

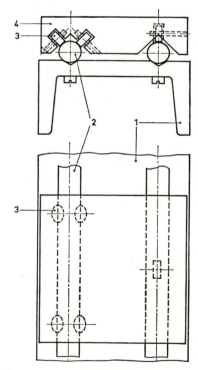

Fig. 3.47. Rod guide with ball bearings.
1 = channel.
2 = round steel bar.
3 = ball bearing.
4 = slide.

Ball and roller guides must be assembled with due regard to the backlash-free retention of the balls, as well as to the exact alignment and parallelism of the track faces (Fig. 3.44).

The stroke or travel S depends on the beam length L and the length of the ball track K (Fig. 3.45). The correct ratio is $S : L \approx 1 : 2$ with $S \leqslant 400$ mm and $S : L_{max} \approx 1 : 1$ with $S > 400$ mm.

The width B of the guide must not exceed the bearing length, that is, the track length K, and stops must be provided at both ends to prevent the slide from running out too far. More familiar types of straight guide, borrowed from conventional mechanical engineering, are used to span longer distances or carry heavier loads. In general, they are constructed on the principle of the lathe bed.

Fig. 3.46 shows a round bar guide constructed on this principle. Two round steel bars are fastened to a channel beam and the slide or carriage, placed on them, is given a vee-groove and a flat guide face. By removing some of the stock or by building it up in places, the mating faces of the slide with the bars can be made to resemble a three-point contact. Instead of the sliding faces, the carriage is sometimes mounted on ball bearings to provide a free-running guide (Fig. 3.47).

The dovetail slide is a time-honoured device in instrument making and mechanical engineering generally. Such slides can be assembled in several ways (Fig. 3.48), the choice being governed by the dimensions, the load and

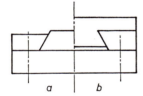

Fig. 3.48. Dovetail slide.
a. Bearing on inner face.
b. Bearing on outer face.

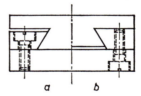

Fig. 3.49. Construction of the dovetail.
a. Screwed from above.
b. Screwed from below.

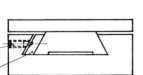

Fig. 3.50. Dovetail with insert strip.
1 = strip or spring.
2 = pointed grub screw.

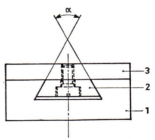

Fig. 3.51. Small dovetail slide.
1 = slotted housing.
2 = inner member.
3 = strip or plate.

the possibilities with regard to mounting. Given accurate finish and correct assembly, the shape of this slide makes it self-retaining. When greased it moves smoothly, without jolting and virtually without play. Final adjustment is effected on assembly by fastening the strips to the plate with cheese-head screws (Fig. 3.49). Another method of adjustment is to insert a strip or spring (Fig. 3.50), which allows for resetting when the slide is in use.

In instrument making, however, there is not always room for these separate strips. The mechanical stage of a microscope is an example of this. Here, the system illustrated in Fig. 3.51 is employed. The member with the dovetail slot is made in one piece, whilst the other guide member is in two parts.

The slide is fitted on assembly by dressing the top face or the slanting side faces of the inner member (2). This process calls for special care and experience on the part of the operator. The top face must be trimmed slightly to eliminate any play, and the two slanting faces trimmed to eliminate any tendency to sticking or heavy-running. The angle between the slanting faces is 60°–80°. The increasing demand for high precision in connection with miniaturization has created a need for very accurate and stable guides. The development of air bearings and fluid bearings will certainly furnish new opportunities for these straight guides, which it would be very difficult, if not impossible, to exploit by the conventional methods.

3.4.2 Bearings (rotary)

Shafts or spindles for revolving optical components, such as mirrors or prisms, or for precise adjustments or measurements of angular displacement, need bearings of a very high standard. Stability and accurate location of the centre of rotation are essential to the proper functioning of the instruments and to the reliability of the results. Imperfections in the bearing members or shaft, such as eccentricity, out-of-roundness, backlash, undue friction, misalignment, vibration and so on affect the precision of the motion adversely.

The choice of bearing construction is governed by the requirements imposed on the motion and the conditions under which the instrument is

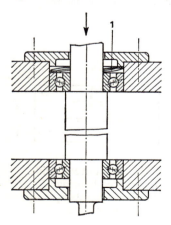

Fig. 3.52. Anti-backlash ball bearing.
1 = spring washer (corrugated or
cupped).

to be used. The bearing of a spindle rotating at high speed in an apparatus for measuring the velocity of light has quite different requirements from that of the prism-table spindle in a spectroscope.

Because ball bearings have minimal friction and are available in high precision quality, they allow very accurate bearing design. Ball bearings are supplied in several different grades. Using bearings of the highest quality, that is, the precision and ultra-precision grades, it is possible to construct mechanisms for minute adjustments of angular rotation, as well as for fast rotary motion.

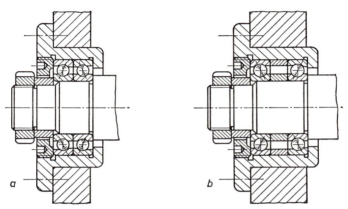

Fig. 3.53. Coupled or paired ball bearing assemblies.
a. Back-to-back in O-arrangement.
b. Face-to-face with spacing rings in X-arrangement.

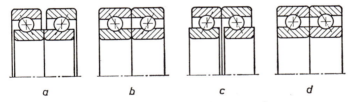

Fig. 3.54. Preloaded pairs of ball bearings.
a. X-arrangement, unloaded.   *c.* O-arrangement, unloaded.
b. X-arrangement, assembled.   *d.* O-arrangement, assembled.

Fig. 3.52 shows the construction of a bearing capable of withstanding a radial force and an axial force in one direction. The spring ensures backlash-free (yet not unduly stiff) retention, whilst compensating any changes of dimension due to temperature fluctuations. Since two matched ball bearings can be supplied, specially designed bearings can be constructed by assembling them in pairs. Fig. 3.53 illustrates two examples of such combined assemblies. The bearings can be mounted without play or even preloaded (Fig. 3.54).

For further particulars, see the extensive tables of dimensions and tolerances given in manufacturers' catalogues, together with explanatory notes concerning assembly and design specifications.

Recent developments in the field of air bearings have been applied to fast rotary motions in optical apparatus, where an air turbine can be used as the prime mover.

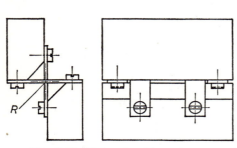

Fig. 3.55. Flat spring hinge.
R = axis of rotation for small
angular displacement.

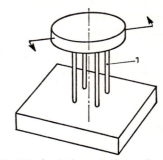

Fig. 3.56. Manipulation platform for
minor movements.
1 = steel wires.

Fig. 3.57. Conical bearing (microscope nosepiece).
1 = journal with top dish.
2 = threaded ring for attachment to tube.
3 = bottom dish.
4 = screw thread for objectives.
5 = lockwasher.

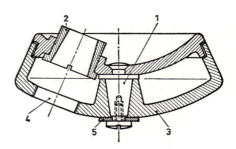

A minor angular displacement corresponding to the rectilinear guide can be obtained with the system illustrated in Fig. 3.55, by means of flat springs. The axis of rotation passes through the points of intersection of the neutral lines of the springs. The platform attached on top of four steel wires, shown in Fig. 3.56, provides a means of manipulation through a small angle. The drive must be supplied by torque alone, since any lateral force would result in linear displacement. Cylindrical or conical sliding bearings can be used where there is no objection to the relatively greater friction involved. The drawing of a microscope nosepiece in Fig. 3.57 shows a conical journal bearing.

3.4.3 Adjusting mechanisms

The choice of mechanical drive to adjust a slide or a turntable depends on the accuracy required and the method employed to measure this movement.

Binoculars, microscopes, cameras, projectors, etc., are adjusted by observing and focusing the image: the size of the movement involved is immaterial. However, provision must be made for very accurate readings of the vertical displacement of a cathetometer telescope, the displacement of a spherometer and the lateral displacement of an eyepiece screw micrometer.

The system of measurement governs the measuring accuracy, so the construction of the drive mechanism is an important factor on the precision of adjustment. Rack-and-pinion drives are used for the adjustment of binoculars, mechanical stages, microscopes with a magnification up to about 250, vertical movements of stands, etc. To get a uniform drive, helical involute teeth are employed, with modules ranging from 0·3 to 1, depending on the force involved. In most cases, the pinion is cut directly on the shaft and is made of steel. The rack is fastened on the slide, and made of brass. Because of the helical teeth, lateral forces are set up in the axial direction and have to be cushioned by a bearing constructed for the purpose. Where the structure permits, the pinion should be wider than the rack and when assembled the pinion projects slightly beyond the rack on both sides This drive is not self-braking and the friction of the slide or the pinion shaft must not be too small. For vertical movements, the friction must be large enough to compensate the weight of the moving member.

By making the friction variable, by clamping the pinion shaft, this force can be employed when the instrument is in use. Where the adjustment involves only a light load, the rack is supported only at its ends. This permits slight deflection of the rack, as a result of which the teeth engage smoothly

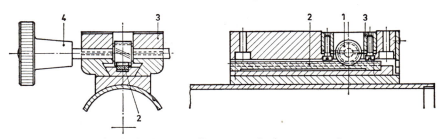

Fig. 3.58. Coarse adjustment of microscope tube.

1 = pinion shaft. 2 = rack. 3 = clamp. 4 = knurled knob.

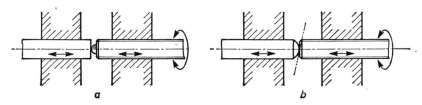

Fig. 3.59. End face of adjusting screw.

a. Spherical (or with embedded ball). *b.* Ground flat.

and without backlash, so that the adjustment takes place evenly and without jolting. Fig. 3.58 illustrates the coarse adjustment of a microscope tube. In this construction, the variable friction of the pinion shaft is brought about by radial clamps at both ends of the shaft. Another method of adjusting the friction is by axial displacement of the knurled knobs.

Adjusting screws are much used in optical instruments, with an intermediate reduction for small precision movements. The different fine adjustments of a microscope are typical, for with these mechanisms, minute movements ranging to < 1 μm are possible.

The end-face of the adjusting screw presses against the driven member and rotates during the adjustment. Fig. 3.59 shows two different shapes. for this contact face: (*a*) spherical and (*b*) flat. The spherical shape causes less deviation, and the alignment of the adjusting screw is less critical than with the other shape. A flat face on the adjusting screw may oscillate if the screw is misaligned, and is therefore more likely to cause deviations.

In instruments where the reading is taken from a drum or knob on the adjusting screw, the position of the nut is important: it should be as close as possible to the end-face (Fig. 3.60). This is particularly necessary for instruments with a calibrated scale on the drum. Placing the nut as close as possible to the end-face minimizes possible deviations caused by changes of temperature and associated dimensional changes.

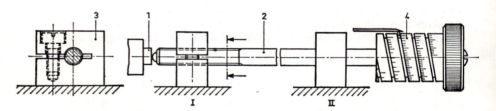

Fig. 3.60. Positioning the nut for calibrated adjustments.

1 = driven arm or pin.   3 = nut. I = correct position; II = wrong position.
2 = adjusting screw.   4 = calibrated drum with index.

3.5. Applied structures

3.5.1 Screw threads

The screw fastenings used to assemble a complete objective or supplementary lens must fit in an optical instrument with a positive clearance. The direction of the optical axis is governed by the screw-on face (Figs. 3.3 and 3.4 earlier). In most cases, the pitch of the screw thread is small, ranging from 0·5 to 1·5 mm for a metric thread, and from 20 to 40 t.p.i. for a Whitworth thread. For cameras with exchangeable objectives, these screw thread sizes are laid down in factory standards. The screw thread for microscope objectives has been standardized internationally (see Fig. 3.61 and DIN standard No. 58888).

3.5.2 *Correcting systems*

The extreme sensitivity of optics to dimensional discrepancies necessitates adjustments during the assembly of precision components with complicated systems. Discrepancies which are within the manufacturing tolerances of the mechanical and optical parts may nevertheless combine to cause undue differences in size on assembly, thereby affecting the quality of the optical system.

However, a correction properly carried out can reduce such defects, even if it does not eliminate them altogether. Fig. 3.62 illustrates the sideways displacement of the correcting system in a microscope objective. During assembly, four screws are inserted in the outer mount. Turning two opposite screws shifts the system in one direction: turning the other pair of screws produces a similar displacement in the same plane but at right-angles. After this adjustment, the lens mounts are pressed firmly together by the threaded ring. The adjusting screws are then removed and the screw holes shut-off by the shield tube.

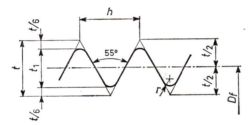

Fig. 3.61. Screw thread for microscope objectives.
Royal Microscopical Society (R.M.S.), with backlash
(effective diameter clearance).
Thread profile W.W. 55°
For 36 t.p.i., pitch $h = 0.7055$ mm
$t = 0.96049\ h = 0.6776$ mm
$r = 0.13733\ h = 0.0969$ mm
$t_1 = 0.64033\ h = 0.4518$ mm

External thread objective	Internal thread tube

Sizes (mm)

$$\left.\begin{array}{l} \text{Db.b.} = 20\cdot274 \\ \text{Dk.b.} = 19\cdot370 \\ \text{Df.b.} = 19\cdot822 \end{array}\right\} \begin{array}{l} +\ 0 \\ -\ 0\cdot075 \end{array} \qquad \left.\begin{array}{l} \text{Db.m.} = 20\cdot320 \\ \text{Dk.m.} = 19\cdot416 \\ \text{Df.m.} = 19\cdot868 \end{array}\right\} \begin{array}{l} -\ 0 \\ +\ 0\cdot075 \end{array}$$

Backlash max. 0·196 mm; min 0·046 mm

Sizes (in.)

External thread	Internal thread
Db.b. = max. 0·7982 in., min. 0·7952 in.	Db.m. = max. 0·8030 in., min. 0·8000 in.
Dk.b. = max. 0·7626 in., min. 0·7596 in.	Dk.m. = max. 0·7674 in., min. 0·7644 in.
Df.b. = max. 0·7804 in., min. 0·7774 in.	Df.m. = max. 0·7852 in., min. 0·7822 in.
Tolerance + 0	Tolerance − 0
− 0·003 in.	' + 0·003 in.

Backlash max. 0·0078 in.; min. 0·018 in.

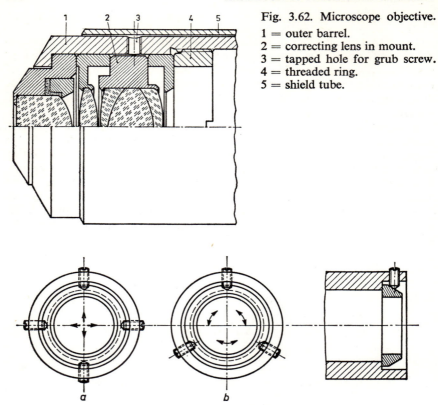

Fig. 3.62. Microscope objective.

1 = outer barrel.
2 = correcting lens in mount.
3 = tapped hole for grub screw.
4 = threaded ring.
5 = shield tube.

Fig. 3.63. Radial adjustment and clamping of mount.
a. With four screws.   *b.* With three screws.

Fig. 3.63 illustrates the use of four screws to adjust and clamp a ring mount containing a cross-hair reticule, a scale, a diaphragm or a lens. Similar adjustments can be made with three screws, but then the motion is not along two axes at right-angles, and is also less readily appraised. A final correction of binocular axes (see Section 3.2.2) is carried out during inspection on the test bench (Fig. 3.2 earlier). In that process, the objectives are moved sideways by means of two eccentric rings (Fig. 3.64). By turning the rings relative to each other, the lenses are adjusted to bring the optical axes of the two binocular tubes parallel. After being locked, the adjusting system is covered completely by the sun visor.

The object of the angular correction of mirror and prism mounts during assembly is to collimate the reflected ray. The angle of adjustment is half the angle through which the ray is to be rotated (refer back to Fig. 3.21). This adjustment corrects the discrepancies caused by the tolerances on the components: the angle involved is relatively small in instruments manufactured to normal standards. Fig. 3.65 shows a method of adjusting a mirror with the aid of two screws. Here, two saw cuts are made to produce a narrow, deformable link, the adjustment being carried out by tightening first one screw, to tilt the mirror slightly, and then the other, to produce a smaller movement in the opposite direction. Precise adjustments of mirrors, etc., in measuring systems often call for a complete mounting fixture fastened on an optical bench. The use of this fixture involves regular adjustments through minute angles (a few seconds of arc), so the optical bench and the fixture must both be extremely stable. Independent adjustments in two directions at right-angles are desirable.

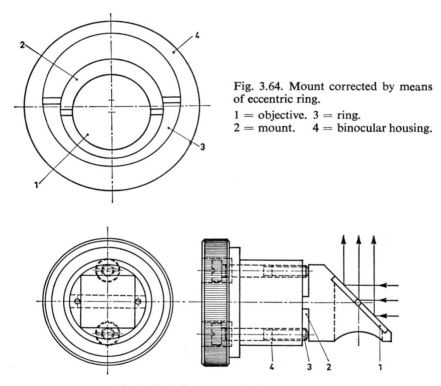

Fig. 3.64. Mount corrected by means of eccentric ring.

1 = objective. 3 = ring.
2 = mount.   4 = binocular housing.

Fig. 3.65. Adjustment of mirror mount.

1 = mirror.   2 = saw cut.   3 = adjusting screw.   4 = centring piece.

Fig. 3.66 illustrates the principle of a mounting fixture in which the optical mount is tilted by two screws. The mount is held under pressure at three points (3, 4, 5) by three springs (2) located at the corners of an equilateral triangle, whose centre of gravity coincides with the axis of the mount. The support points are at the corners of a right-angled isosceles triangle whose centre of gravity likewise coincides with the mount axis. The fixed support point is at the corner of the right-angle and the two adjustable points are at the acute corners of this triangle. Adjustment of one of the points causes

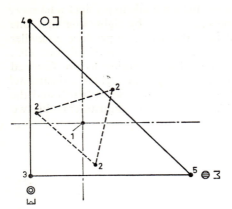

Fig. 3.66. Diagram of adjustment of precision mounting fixture.   •

1 = axis of optical mount.
2 = springs at corners of equilateral triangle.
3 = fixed support point (hollow).
4 = movable point for horizontal adjustment (plane).
5 = movable point for vertical adjustment (groove).

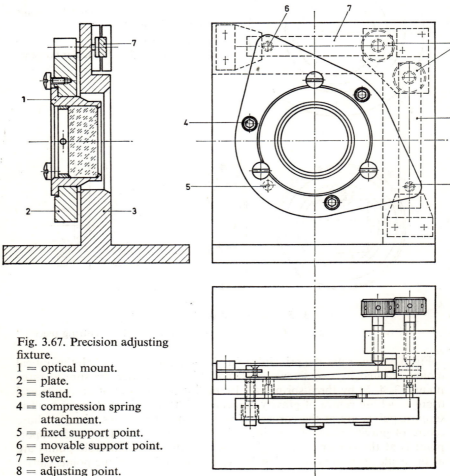

Fig. 3.67. Precision adjusting fixture.
1 = optical mount.
2 = plate.
3 = stand.
4 = compression spring attachment.
5 = fixed support point.
6 = movable support point.
7 = lever.
8 = adjusting point.

the mount to tilt about the opposite side of the right-angle. Figure 3.67 shows the construction of this fixture.

3.5.3 Lens units

The lens unit system, consisting of several lenses, has to be assembled into one complete optical component. This component is in turn part of an instrument which can be made to meet different requirements by changing the lens unit as a whole. Such instruments or measuring systems can therefore be used for a wide range of work. In microscopy, for example, the magnification and the mode of observation can be suited to the object, by choosing the appropriate objective and eyepiece. In photography, exchangeable objectives give scope for more universal use of the camera. One or two examples of application and construction taken from the wide variety of objectives, eyepieces, condensers and so on will now be discussed.

(a) Stack mounting

A number of lenses can be fastened together in a unit by making a single outer barrel in which all the lenses are mounted and clamped axially by means of a single fixing part. By this method, a compact and accurate component is obtained.[5, 6]

There are three possibilities, as follows:

The lens diameters are identical.

The lens diameters are different.

Each lens is fastened in a separate metal ring.

Fig. 3.68 is an example of lenses of equal diameter. Centring is accomplished by grinding all the lenses to the same size, to fit in the outer barrel. The metal spacing rings keep the lenses at the correct distance from one another, whilst the threaded ring clamps this stack of components firmly together.

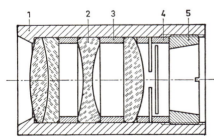

Fig. 3.68. Stack mounting (lenses of the same diameter).

1 = barrel.   4 = spring washer.
2 = lens.   5 = threaded ring.
3 = spacer ring.

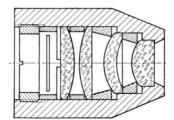

Fig. 3.69. Stack mounting (lenses of different diameters).

With these units, the risk of undue stress in the glass, due to changes of temperature, is considerable, but these forces can be compensated by fitting a resilient member.

Figure 3.69 shows a similar structure, but with lenses of different diameters.

The microscope objective in Fig. 3.70 is an example of a structure with lenses set in separate rings. It involves little risk of stress in the lenses, so this structure is very suitable for the manufacture of strain-free objectives, provided the metal parts are stress-relieved.

With a powerful objective, the free working distance is exceptionally short ($\ll$ 1 mm). Hence, there is considerable risk of accidental damage to the specimen because the objective is screwed down too far, particularly where the fine adjustment operates through a reduction gear, so that it is impossible to feel when the objective touches the object.

Figure 3.71 shows a spring-loaded mount, which lessens the risk of broken or damaged lenses.

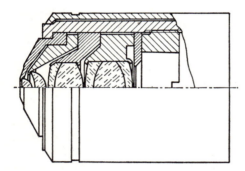

Fig. 3.70. Stack mount (lenses mounted individually).
Powerful microscope objective.

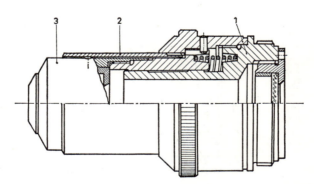

Fig. 3.71. Microscope objective with spring-loaded mount.
1 = fixed top piece with guide.   3 = spring-loaded bottom piece.
2 = protective barrel.

(b) *Screw mounts*

In many systems, the lenses or lens mounts have to be fastened in the lens unit independently. The measuring eyepiece shown in Fig. 3.30 earlier must be partially taken down in order to change the graticule. In the camera objectives of Figs. 3.72 and 3.73, the shutter and the iris diaphragm are installed between the lenses. This screw mount construction, combined with stack mounting where necessary, is needed for units containing lenses differing considerably in diameter, for complex lens shapes, units incorporating diaphragm and shutter, etc.

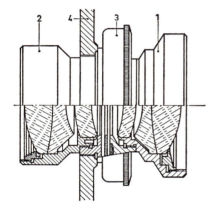

Fig. 3.72. Camera objective (Schneider-Symnar).

1 = front piece of objective.
2 = back piece of objective.
3 = shutter and diaphragm.
4 = camera.

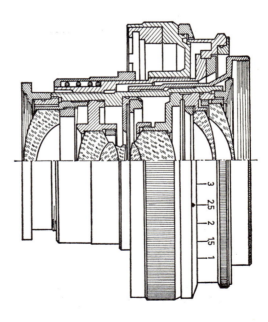

Fig. 3.73. Camera objective (Zeiss-Biogon.).

(c) *Compensation rings*

An elastic ring or bush is used to compensate for changes of length in stack mounts, or to prevent undue compressive forces on lenses as the screw fastening is tightened.

Fig. 3.74 shows a ring with three saw cuts in one plane, and three contact points in the pressure plane.

Two sets of three saw cuts are made to obtain the elastic bush shown in Fig. 3.75.

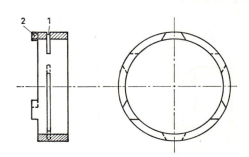

Fig. 3.74. Elastic ring.
1 = saw cut.   2 = pressure face.

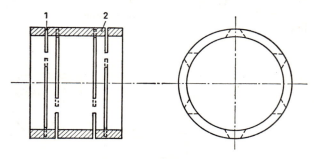

Fig. 3.75. Elastic bush.
1 = saw cut.   2 = elastic spacer.

3.5.4 *Eliminating reflection*[2]

Special care must be taken to reduce or counteract the adverse effects of stray light rays reflected by the insides of tubes and rings. These rays, reflected at the inner faces of metal members, affect the quality of the image seriously, and make it difficult, or impossible, to observe the object.

Rays which, for various reasons, are deflected outside the beam must not be reflected back into the optical system. The inside diameters of tubes or other metal parts linking two sections of an optical system must be made large enough: the marginal rays must not be cut or interrupted. Such metal surfaces are dulled or blackened to make them less reflecting and can be given a matt finish by sand blasting, sanding, etching or grooving. The grooves may be helical or in the form of separate rings.

Another kind of dull finish can be obtained by painting the surfaces matt black or by chemical processing, but matt black surfaces are not mechanically strong and are most sensitive to grease and dirt. Extra stops inserted between the optical parts prevent oblique rays from being reflected back into the beam (Fig. 3.76).

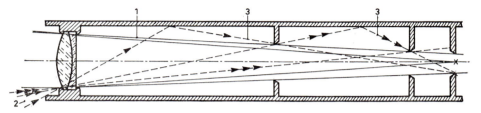

Fig. 3.76. Positioning of intermediate stops.

1 = marginal ray.   2 = ray entering at an angle.   3 = reflected ray.

A diaphragm or tube placed in front of the optical system will keep out rays incident at an angle. The lens hoods of cameras and binoculars fulfil this purpose, and also protect the lens against impact, and against rain and dirt from the outside.

A close-fitting eyeguard on the eyepiece prevents dazzling light reflections reaching the eye lens from behind the observer's head. Such eyeguards are particularly important in viewing with both eyes, as with binocular field-glasses and binocular microscopes. The eyeguard should be mounted to rotate on the eyepiece and should fit closely round the eye. A rubber eye-guard moulds readily to the eye, and is less vulnerable to accidental impact, etc.

REFERENCES

[1] M. POLLERMANN, *Bauelementen der physikalischen Technik*, Springer Verlag, Berlin, 1955.
[2] H. NAUMANN, *Optik für Konstrukteure*, Wilhelm Knapp Verlag, Düsseldorf H., 1970.
[3] B. K. JOHNSON, *Optics and Optical Instruments*, Dover, New York, 1960.
[4] W. J. SMITH, *Modern Optical Engineering*, McGraw-Hill, New York, 1966.
[5] R. KINGSLAKE, *Applied Optics and Optical Engineering*, Academic Press, New York, 1965/67.
[6] K. H. SIEKER, *Taschenbuch der Feinwerktechnik*, Wintersche Verlashandlung, 1965.
[7] K. MÜTZE, *ABC der Optik*, F. A. Brockhaus Verlag, Leipzig, 1961.
[8] G. SCHULZE, *Das Zentrieren sphärischer Linsen durch mechanisches Ausrichten mittels Spannrohren*, Feinwerktechnik, August 1964, p. 307.
    W. ZSCHUMMLER, *Feinoptik-Glasbearbeiting*, Carl Hanser Verlag, Munich, 1963.
    A. JACOBSEN and W. RIMKUS, *Faseroptik—Eigenschaften und Anwendungen*, Feinwerktechnik, March 1967, p. 111.

G. Zeichen, *Feinwerktechnische Beziehungen zwischen optischer und optisch-elektronischer Messtechnik und neuen Fertigungsmethoden*, Feinwerktechnik, April 1967, p. 153.

K. Heinecke, *Photoelektrisches Einfangen vorn Strichmarken*, Feinwerktechnik, April 1967, p. 160.

Index

HANDBOOK OF PRECISION ENGINEERING

HANDBOOK OF PRECISION ENGINEERING